K. Porter

# **Antibodies**
# Volume I

# a practical approach

# TITLES PUBLISHED IN THE PRACTICAL APPROACH SERIES

Series editors:

**Dr D Rickwood**
Department of Biology, University of Essex
Wivenhoe Park, Colchester, Essex C04 3SQ, UK

**Dr B D Hames**
Department of Biochemistry, University of Leeds
Leeds LS2 9JT, UK

Affinity chromatography

Animal cell culture

Biochemical toxicology

Biological membranes

Carbohydrate analysis

Centrifugation (2nd Edition)

DNA cloning

Drosophila

Electron microscopy in molecular biology

Gel electrophoresis of nucleic acids

Gel electrophoresis of proteins

HPLC of small molecules

Human cytogenetics

Human genetic diseases

Immobilised cells and enzymes

Iodinated density gradient media

Lymphocytes

Lymphokines and interferons

Mammalian development

Microcomputers in biology

Mitochondria

Mutagenicity testing

Neurochemistry

Nucleic acid and protein sequence analysis

Nucleic acid hybridisation

Oligonucleotide synthesis

Photosynthesis: energy transduction

Plant cell culture

Plasmids

Prostaglandins and related substances

Spectrophotometry and spectrofluorimetry

Steroid hormones

Teratocarcinomas and embryonic stem cells

Transcription and translation

Virology

Yeast

# Antibodies Volume I

# a practical approach

---

Edited by

**D Catty**

Department of Immunology, University of Birmingham Medical School,
Vincent Drive, Birmingham B15 2TJ, UK

OXFORD · WASHINGTON DC

IRL Press
Eynsham
Oxford
England



First Published 1988



British Library Cataloguing in Publication Data

Antibodies.
Vol. 1: A practical approach
1. Organisms. Antibodies
I. Catty, D. II. Series
574.2′93

ISBN 0 947946 86 1 (hardbound)
ISBN 0 947946 85 3 (softbound)

Typeset by Infotype and printed by Information Printing Ltd, Oxford, England.

# Preface

Under the title *Antibodies — A Practical Approach*, of which this volume is the first of two, IRL Press is extending its highly successful Practical Approach series of texts in the biosciences into the sphere of immunological methods. The specificity and antigen-binding properties of antibodies have been exploited since the start of the century but the last 20 years have seen an enormous growth in their application. There can be few laboratories occupied in the study of living systems and cell products at the physiological and biochemical level that have not come to appreciate the value of antibodies as the preferred and often essential tools to identify, quantify and probe the structure and biological properties of antigenic molecules in the context of their own work. The range of applications is enormous — animal and plant hormones, enzymes, cell receptors and differentiation markers, serum proteins, tissue and cell-specific antigens, tumour-associated antigens, microbial and parasite antigens and many, many others. To be effective across such a diverse range of applications the use of antibodies relies upon two closely linked areas of expertise — that of producing highly specific antibody reagents that perform reliably, and that of choosing and performing reliable antibody assay procedures appropriate to the task. This book has brought together these two aspects of antibody methodology; linking how to produce and quality control antibodies with how to use them. I have drawn heavily upon the expertise and long experience of several of my colleagues in the Department of Immunology in Birmingham but where necessary have included contributions from experts of other centres to achieve a broader perspective. The result is a practical book in two volumes which will, I hope, be of lasting value to those many scientists who need to produce either polyclonal or monoclonal antibodies, and to the larger number of scientists, students and technicians who may need guidance on antibody methods. As intended in a practical approach series the chapters on antibody methods adopt a bench manual format with the aim of being of most value as bench aids. The methods described are of three categories — those that are used routinely and have wide application, such as reactions in gels, haem-agglutination, ELISA and radioimmunoassays, immunofluorescence and immuno-peroxidase methods and affinity purification of molecules; those of more restricted application but of major importance in the clinical field, such as red cell typing and tissue typing, the technologies of which continue to advance with a need for contemporary appraisal of methods; and recently developed methods, such as fluorescence activated cell sorting and the use of antibodies in tumour imaging, that are of growing importance and clearly have great future potential.

D.Catty

# Contributors

J.Arvieux and A.F.Williams
*MRC Cellular Immunology Unit, Sir William Dunn School of Pathology, University of Oxford, South Parks Road, Oxford OX1 3RE, UK*

G.Brown, D.Catty, J.Gordon, N.R.Ling and C.Raykundalia
*Department of Immunology, University of Birmingham Medical School, Vincent Drive, Birmingham B15 2TJ, UK*

# Contents

# Abbreviations

| | |
|---|---|
| ACD | acid citrate dextrose |
| AEC | 3-amino, 4-ethylcarbazole |
| AECM | aminoethyl-carboxymethyl |
| AIHA | auto-immune haemolytic anaemia |
| AS | ankylosing spondylitis |
| BCG | Bacille Calmette-Guérin |
| BDB | *bis*-diazitized benzidine |
| BGG | bovine $\gamma$ globulin |
| BIS | *N,N'*-methylene bisacrylamide |
| BSA | bovine serum albumin |
| BSS | balanced salt solution |
| C-CIEP | counter-current immunoelectrophoresis |
| CD | cluster-designation |
| CDC | complement-dependent cytotoxicity |
| CFDA | carboxy-fluorescein diacetate |
| CFT | complement fixation test |
| CLL | chronic lymphocytic leukaemia |
| CREGS | cross-reactive groups |
| DAB | diaminobenzidine |
| DAGT | direct antiglobulin test |
| 2D-IEP | two-dimensional immunoelectrophoresis |
| DMSO | dimethyl sulphoxide |
| DNFB | 2,4 dinitrofluorobenzene |
| DNP | 2,4 dinitrophenol |
| DTT | dithiothreitol |
| EBV | Epstein—Barr virus |
| EIA | enzyme immunoassays |
| ELISA | enzyme-linked immunosorbent assay |
| EMIT | enzyme-modulated immunotest |
| FACS | fluorescence-activated cell sorter |
| FCA | Freund's Complete Adjuvant |
| FCS | fetal calf serum |
| FDA | fluorescein diacetate |
| FIA | Freund's Incomplete Adjuvant |
| GDD | gel double diffusion |
| HAI | haemagglutination inhibition |
| HAT | hypoxanthine aminopterin and thymidine |
| HD | haemolytic diluent |
| HGPRT | hypoxanthine—guanine phosphoribosyltransferase |
| HLA | Human Leukocyte series A |
| HRP | horseradish peroxidase |
| HRPMI | Hepes-RPMI |
| HSA | human serum albumin |
| IAGT | indirect antiglobulin test |
| IEF | isoelectric focusing |
| IEP | immunoelectrophoresis |
| IFT | immunofluorescence test |

| | |
|---|---|
| Ig | immunoglobulin |
| IPT | immunoperoxidase test |
| IRMA | immunoradiometric assay |
| KLH | Keyhole limpet haemocyanin |
| LEE | low electroendosmosis |
| LISS | low ionic strength saline |
| LPS | lipopolysaccharide |
| Mab | monoclonal antibodies |
| MAD | minimum agglutining dose |
| MAHD | minimum antibody haemolytic dose |
| MDP | muramyl dipeptide |
| 2-ME | 2-mercaptoethanol |
| MSH | *Maia squinada* haemocyanin |
| Myo | myoglobin |
| OVA | ovalbumin |
| PAGE | polyacrylamide gel electrophoresis |
| PBMC | peripheral blood mononuclear cells |
| PBS | phosphate-buffered saline |
| PEG | polyethylene glycol |
| PFCDC | platelet fluorescence complement-dependent cytotoxicity |
| PHA | passive haemagglutination |
| PLL | poly-L-lysine |
| PMSF | phenyl methyl sulphonyl fluoride |
| PPD | purified protein derivative |
| PRP | platelet-rich plasma |
| PT | porcine thyroglobulin |
| RIA | radioimmunoassay |
| RID | radial immunodiffusion |
| RIEP | rocket immunoelectrophoresis |
| RPHA | reverse passive haemagglutination |
| RRID | reverse radial immunodiffusion |
| SDS | sodium dodecyl sulphate |
| sIg | surface immunoglobulin |
| SPDP | *N*-succinimidyl-3(2-pyridyldithio)proprionate |
| SRBC | sheep red blood cells |
| Tc | cytotoxic T cells |
| TEMED | *N,N,N′,N′*-tetramethylenediamine |
| TNBS | trinitrobenzene (picryl) sulphonic acid |
| TNP | trinitrophenol |
| WB | Western blotting |

# Introduction

DAVID CATTY

In 1890 von Behring and Kitasato (1) prepared bacterial antitoxins and demonstrated their neutralizing properties. This discovery set in motion an era of serological studies remarkable for its achievements in revealing both the properties of antibodies and complement in bacterial infections and the diagnostic value of antisera in, for instance, typhoid (Widal Test, 1896) and syphilis (Wasserman Test, 1906). The flourishing contemporary practice of serodiagnosis, as a branch of microbiology, follows the traditions of these early discoveries. As antibodies are evolved as a principle means of defence against infection it might reasonably have been assumed that their relevance would rest exclusively in this realm, with applications restricted to determining the infection response and the antigens involved. However, other rich avenues of exploration were soon to reveal that the antibody response had an apparently limitless range of specificity which could be used to define, isolate and measure a universe of immunogenic molecules not confined to the products of microorganisms. Indeed most complex molecules with molecular weights greater than 5000 that are not intrinsic to the species immunized were found to be immunogenic, and Landsteiner in 1936 (2) showed that smaller molecules (haptens) chemically coupled to a larger carrier could elicit antibodies of exquisite specificity.

The foundations of immunoassay as an empirical method for antigen study rest firmly on the critical understanding of the nature of antibody specificity thus defined in principle, and on an appreciation of the quantitative nature of antigen–antibody interaction. The ability of antisera to precipitate antigens from solution, recorded by Kraus as early as 1897 (3) with *Vibrio cholerae* culture fluid, was shown to be a quantitative phenomenon by Heidelberger in 1939 (4) who developed the quantitative precipitation test and purified antibody for the first time. Precipitation of antigen, which is the basis of many contemporary forms of immunoassay, depends upon the multi-determinant specificity of antisera whose constituent antibodies bind to the mosaic of determinants on complex antigens, constructing a lattice of combined molecules which, at appropriate relative concentrations of antigen and antibody, come out of solution. Although Marrack (5) had laid down the theoretical basis of the reaction in 1938, its exploitation in simple immunoassays that could measure single antigen systems was not possible until Oudin (6) in 1946 demonstrated that diffusion of antibodies and antigens in the fluid phase across the interface of opposing agar gels could result in quantitative assessment to both reacting elements by the position of the resulting precipitation bands. Diffusing in a gradient of concentration into each other the antibody–antigen systems formed visible complexes where their relative proportions are conducive to precipitation. This technological breakthrough led rapidly to the application of double immunodiffusion in gel developed by both Elek (7) and Ouchterlony (8) in 1948, by which, for the first time, antigenic relationships between molecules would be easily determined and the specificity of antisera assessed. Immunoelectrophoresis was developed by Grabar and Williams in 1953 (9) and radial immunodiffusion with its quantitative properties by Mancini, Carbonara and Heremanns in 1965 (10).

The principle of combining electrophoretic migration of antigen in agar with precipitation by antibody was further exploited by Laurell (11) in 1966 in developing 'rocket' immunoelectrophoresis for the rapid estimation of single antigens from mixtures using unispecific antibody in the gel, and two-dimensional (crossed) immunoelectrophoresis which allows both qualitative and quantitative assessment of multiple components of antigen mixtures by using multispecific antisera.

Whilst gel precipitation methods occupy an important position in the study and measurement of antigens, and in quality control of specificity and titre of precipitating antibodies, the limits of sensitivity are about 5 $\mu$g/ml for antigen and they cannot be applied to the measurement of low molecular weight antigens that form only insoluble complexes. For this reason alternative methods were needed to measure some antigens. In 1945 Coombs *et al.* (12) described the antiglobulin reaction in which anti-human immunoglobulin sera agglutinated human erythrocytes from certain subjects on whose cells antibodies had previously bound. In addition to developing the system as an important clinical test (Coombs' Antiglobulin Test), Coombs pioneered the use of red cells as an indicator system in immunoassays. With the discovery of reliable means to attach antigens to the red cell membrane, sensitive haemagglutination assays of wide application were soon developed. Subsequently it was shown that agglutination of antigen-coated cells by antibody could be inhibited by the presence of trace amounts of competing unfixed antigen and this provided a novel and very sensitive method of quantifying antigen in solution. Other technical advances were to follow, including reverse agglutination with antibody-coated red cells.

A further milestone in sensitive assays came with the discovery of radioisotopic labelling techniques for antigens, antibodies and small peptide hormones by Farr in 1958 (13), and Yalow and Berson in 1960 (14). The advent of radioimmunoassays allowed the measurement of small molecules such as insulin to extremely low concentrations which extend down to the pmol/litre range. Such assays made special demands on the technology of antiserum production using hapten (hormone) conjugates on carrier molecules for animal immunization to produce a degree of specificity and binding affinity of the antibody not previously required.

The use of a coupled label to provide an indirect signal of antigen–antibody interaction had first been adopted by Coons and Kaplan in 1950 (15) when they used coupled fluorescein as a light-emitting dye in the UV spectrum to investigate antibody production and antigen localization in lymphoid tissue. Immunocytochemistry now utilizes both fluorescent dyes and, increasingly, enzyme/substrate systems as in the immunoperoxidase method. The latter arose from the demonstration by Avrameas and Uriel (16) and by Nakane and Pierce (17) in 1966, that enzymes could also be coupled to antibody or antigen. The importance of this discovery is reflected in the now widespread application of chromogenic, fluorogenic or luminescent enzyme/substrate interaction signals for the detection and measurement of soluble antigens, with an attained sensitivity that approaches that of a radioimmunoassay. This was pioneered by Engvall and Perlmann (18) in 1971 and Van Weeman and Schuurs in the same year (19) and today is translated into an enormous variety of solid phase enzyme-based systems used both in research

**Table 1.** Some major applications of prepared antibodies.

| | |
|---|---|
| 1. | *Identification and study of molecules* |
| | Detection of individual antigens in complex mixtures in solution. |
| | Testing purity of separated molecules. |
| | Determining sites of production, expression, deposition and activity of molecules in or on cells and tissues. |
| | Determining structure/function correlates of molecules such as enzymes, antibodies, hormones, cytokines and cell receptors. |
| | Definition of differentiation and oncofetal antigens on cells. |
| | Definition of cell populations including tumours. |
| | Definition of allelic antigens of single genetic loci—e.g. Major Histocompatibility Complex. |
| | Determining antigenic relationships between molecules—e.g. of different species, between hormones etc. |
| | Lymphocyte (HLA) typing. |
| | Red cell (blood group) typing. |
| | Probing for recombinant DNA (antigen) expression in clones of *Escherichia coli*, yeast etc. |
| | Antigen diagnosis of parasitic, fungal, bacterial and viral infections by detection of organisms or their products in blood, sputum, urine, faeces, swabs, CSF etc. |
| | Serotyping and subtyping of bacteria and viruses. |
| | Definition of antigen variants and gene conversion antigens of parasites. |
| | Detection of infectious agents or their antigens in vectors and reservoir hosts. |
| 2. | *Measurement of molecules* |
| | Blood components as disease markers—e.g. altered levels or composition of immunoglobulins, complement, immune complexes, acute phase proteins, cardiac myosin, alphafetoprotein in neural tube defects. Presence of cancer (carcinoembryonic) antigen (CEA) etc. |
| | Hormones:endocrinopathies, oestrus cycle, pregnancy etc. |
| | Specific antibodies: infection, immunity status, autoimmunity, antibody response to allergens etc. |
| | Mediators of cellular activity: interferons, interleukins, growth factors, chemotactic factors etc. |
| | Drugs and toxins. |
| 3. | *Purification of molecules and cells* |
| | Antibody affinity purification methods for antigens. |
| | Precipitation of an RNA/antigen complex from cytosol in construction of enriched gene libraries for the synthesis of antigen by recombinant DNA technology. |
| | Antibody labelling of cells for their separation by Fluorescence Activated Cell Sorting. |
| | Binding to discrete cell membrane components for their subsequent isolation. |
| 4. | *Real and potential therapeutic applications* |
| | Immunosuppression—e.g. antilymphocyte globulin; antibodies to discrete lymphocyte subpopulations; anti-idiotype antibodies. Antibodies to other cell receptors. |
| | Antibodies to tumour cell antigens, tumour imaging and cytotoxicity. |
| | Anti-idiotype antibodies as 'internal image' immunogens. |
| | Antibodies to bacterial toxins, bacterial and viral, parasite and fungal somatic antigens. |
| | Anti-human chorionic gonadotropin hormone and anti-trophoblast antibodies in fertility control. |

and in routine serological tests. The nature of these tests come closest, with present available technology, to fulfilling many of the ideal properties of an immunoassay—they are simple to perform, inexpensive, especially with home-produced reagents, reliable, robust and attain levels of sensitivity that cover most requirements. Their application has been greatly extended by the use of monoclonal antibodies.

The steady growth of experience in making specific antisera and in absorbing them to refine their capacity to recognize individual antigens in complex mixtures saw major new applications evolving in the study of cell membrane components and this again has been greatly enhanced lately by the use of monoclonal antibodies. Complement-fixing antibodies were found to be able to eliminate, selectively, target cells in mixed suspensions and cultures, leaving as a viable residue other cells not bearing the specific antigen. Apart from the obvious direct clinical applications of such antibodies in, for instance, immunosuppression with antilymphocyte serum, cytotoxic antibodies have been used as a major laboratory method for tissue typing since the discovery of lymphocytotoxic antibodies in mice by Gorer and O'Gorman in 1956 (10). These were applied to man after the discovery of the HLA systems of antigens by Dausset in 1958 (21); the methods were modified by Terasaki and McClelland in 1964 (22). Antibodies are also increasingly used to label cell populations defined by possession of discrete surface markers. This is a rapidly developing field of application which has been greatly facilitated by the invention in the 1970s of the Fluorescence Activated Cell Sorter by which cell populations, labelled by fluorescent antibody, can be enumerated and separated by fluorescence intensity. Again the advent of monoclonal antibodies has extended the scope of definition of cell populations by their surface molecular markers using this automated technique.

The value of antibodies to cell surface components rests not only in their power to destroy selectively, or separate, cell populations but also in their ability to define the discrete target molecules. Assays for components of the cell surface are now available which utilize antibodies for binding to, and separating, membrane molecules, and for revealing their nature through affinity labelling of discrete entities of membrane eluates separated on polyacrylamide gels. These methods are at the forefront of contemporary research in many areas of cell biology.

It is almost a century since antibodies were discovered in the context of bacterial infections. Their applications have in the intervening period revolutionized our ability to detect and measure molecules and determine their properties (see *Table 1*).

This book sets out logically the methods required for the preparation and application of antibodies. Because the field is so large, and the pace of development of new applications and technologies is very rapid, it is inevitable that some aspects will have been neglected. We have concentrated on providing protocols for those methods that are in popular use, have the widest application and teach effectively the principles. The second aim is to cover those methods of singular importance in applied research, and others which, because of their clinical significance in areas of rapidly expanding knowledge, require an appraisal of contemporary laboratory practice.

## REFERENCES


1. von Behring,E. and Kitasato,S. (1964). In *Immunology for Students of Medicine*. Humphrey,J.H. and White,R.G. (eds), Blackwell Scientific Publications, Oxford, 2nd edition p. 5.
2. Landsteiner,K. (1936) *The Specificity of Serological Reactions*. Charles C.Thomas [reprinted by Doner Press, New York, 1962].
3. Kraus,R. (1897) *Wien Klin. Wochenschr.*, 10, 736.
4. Heidelberger,M. (1939) *Bacteriol. Rev.*, **3**, 49.
5. Marrack,J.R. (1938) *Medical Research Council Spec. Rep. Ser.*, No. **230**.
6. Oudin,J. (1946) *C. R. Acad. Sci., Paris,* **222**, 115.
7. Elek,S.D. (1948) *Br. Med. J.*, **1**, 493.
8. Ouchterlony,O. (1948) *Arkiv für Kemi, Mineralogi och Geologi, Bol.*, **26B**(14), 1.
9. Grabar,P. and Williams,C.A. (1953) *Biochim. Biophys. Acta,* **10**, 193.

10. Mancini,G., Carbonara,A.O. and Heremans,J.F. (1965) *Immunochemistry,* **2**, 235.
11. Laurell,C.-B. (1966) *Anal. Biochem.,* **15**, 45.
12. Coombs,R.R.A., Mourant,A.E. and Race,R.R. (1945) *Br. J. Exp. Pathol.,* **26**, 255.
13. Farr,R.S. (1958) *J. Infect. Dis.,* **103**, 239.
14. Yalow,R.S. and Berson,S.A. (1960) *Clin. Invest.,* **39**, 1157.
15. Coons,A.H. and Kaplan,M.A. (1950) *J. Exp. Med.,* **19**, 1.
16. Avrameas,A. and Uriel,J. (1966) *C. R. Acad. Sci., Paris,* **262**, 2543.
17. Nakane,P.K. and Pierce,G.B. (1966) *J. Histochem. Cytochem.,* **14**, 929.
18. Engvall,E. and Perlmann,P. (1971) *Immunochemistry,* **8**, 871.
19. Van Weeman,B.K. and Schuurs,A.H.W.M. (1971) *FEBS Lett.,* **15**, 232.
20. Gorer,P.A. and O'Gorman,P. (1965) *Transplant. Bull.,* **3**, 142.
21. Dausset,J. (1958) *Acta Haematol.,* (Basel), **20**, 156.
22. Terasaki,P.I. and McClelland,J.D. (1964) *Nature,* **204**, 998.

CHAPTER 1

# Properties of antibodies and antigens

DAVID CATTY

## 1. ANTIGENS AND IMMUNOGENS

The traditional meaning of antigen is any molecule which evokes a specific immune response—whether this is antibody production, cell-mediated immunity or tolerance. It is also used more broadly to denote mixtures of molecules, whole microorganisms or cells used as an immunizing entity or as a complex target for antibody binding in immunoassays; thus red cells may be described as an immunizing antigen or as the antigen in agglutination tests. To distinguish between molecules that evoke antibody production (as well as cellular immunity) and those that are the targets for antibody binding, it is the modern convention to use the term *immunogen* for the former and *antigen* for the latter. This is helpful in partitioning concepts for immunogenicity from antigenic properties of molecules related exclusively to antibody binding. To be an immunogen a molecule must possess a degree of intrinsic structural complexity (immunogenicity). Natural immunogens are usually macromolecules of protein or carbohydrate composition, or combinations of these (with or without lipids that are themselves non-immunogenic), with a molecular weight greater than 1000 and normally above 5000. Highly immunogenic molecules are those usually of greater than 100 000 mol. wt. Synthetic polypeptides and other copolymers may be immunogens if they possess these properties. Smaller structures such as substituted aromatic groups, steroids and peptides can induce specific responses if first covalently coupled to larger carrier backbones where they act as haptens on constructed immunogens. Immunogenicity is also influenced by the relatedness (foreignness) of the molecule to the immunized species and in this context is defined by the recipient's immune system. Natural molecules of closely-related species have low immunogenicity. In addition individuals of any species vary in responsiveness to discrete immunogens. This has a genetic basis and is most critically observed in the responses of inbred strains and with immunogens of restricted complexity.

In terms of immunoassays, antigen is best defined in the restricted sense of a molecule that binds specifically to antibody, by virtue of possessing antigen(ic) determinants or epitopes (see Section 2). Antigens can be non-immunogenic subunits of the immunogen, free haptens, haptens attached to different carriers, or natural molecules sharing some determinants with the immunogen. The latter is the basis of antigenic cross-reactivity (see Section 5).

## 2. ANTIGEN(IC) DETERMINANTS: EPITOPES

These are structurally-defined sites of three-dimensional composition on both immunogens and antigens. They are the sites of engagement with lymphocyte-specific receptors in induction of immune responses and the sites on antigens to which specific antibodies bind by their complementary combining sites. Complex antigens possess a mosaic of different antigen determinants to which a multiplicity of antibody specificities may be present in an antiserum as a result of polyclonal stimulation. Each determinant has a size corresponding to 4–6 amino acid or sugar residues, but the combination of shapes and charge distributions that can be offered by determinants is nevertheless very large. The potentials of the antibody response are sufficiently diverse to complement determinant diversity. Indeed, even at the single determinant level, combining antibodies may be clonally heterogeneous with a range of individual affinities (see Section 4).

## 3. ANTIBODY COMBINING SITES (ANTIGEN BINDING SITES)

These are the sites on antibody molecules which have specific binding activity for antigen determinants. They are constructed from the folding of variable domains of light and heavy chains of antibody molecules to form three-dimensional binding spaces with an internal surface shape and charge distribution complementary to these features of antigen determinants. Combining sites donate specificity to the antibody molecule. It is estimated that the diversity-generating potential of the immune system offers in the range of $10^6 - 10^7$ different specificities to the antibody response.

In interaction with macromolecular antigen, different classes of antibody (and IgG subclasses) possess one or more functional binding sites. Most forms of IgG (the predominant serum antibody following chronic immunization of animals) has two combining sites that can 'bridge' between two antigen molecules. IgM antibodies have an effective valency of five and are therefore very efficient in forming precipitates with antigen and in agglutinating cells.

## 4. ANTIBODY AFFINITY AND AVIDITY

Operationally, antibody affinity relates to the exactitude of stereochemical fit of an antibody combining site to its complementary antigen determinant. This can be considered in thermodynamic terms as the strength (energy) of close range non-covalent force interactions between receptor (combining site) and ligand (determinant). Mathematically, affinity is expressed as an association constant ($K$, l/mol) which may be calculated under conditions where equilibrium can be achieved between ligand bound and unbound in its reversible interaction with a homogeneous source of combining site. As most antibody preparations are heterogeneous in affinity towards a single antigen determinant, the derived association constant represents an average value. Antisera of multiple specificity, that is, specific to many determinants on an antigen, cannot be assessed for affinity. They can, however, be compared one with another in overall binding characteristics to antigen—the overall 'strength' of the reaction, in the chosen assay system, denotes the so-called avidity of the antibodies. Avidity is a strictly functional term that defines the performance of antibodies (antiserum) in a particular test. Under the conditions of the test the multiple interacting components are each assumed to obey the laws of mass action as applicable to the formula: $\mathrm{Ab} + \mathrm{Ag} \rightleftharpoons \mathrm{Ab.Ag}$. With this assumption, avidity can be determined and expressed as an association constant $K_a$.

## 5. ANTIBODY SPECIFICITY: CROSS-REACTIVITY

Each antibody molecule is specific to the antigen determinant to which it binds—this represents the most closely-defined level of specificity. The specificity of an antiserum, by contrast, reflects the many specificities of the constituent antibodies. An antiserum may bind exclusively to one antigen if the range of its constituent antibody specificities extends to determinants exclusive to that antigen. Commonly, however, some antigen determinants are shared between molecules—especially if these are similar molecules of related species. In this case some antibodies induced in response to one antigen may bind to another antigen and are then said to be cross-reacting and the containing antiserum is said to lack specificity or be cross-reactive. Cross-reactive antisera can usually be rendered specific to the inducing antigen by a process of absorption, by which cross-reacting antibody components are removed by exposure to antigens with shared determinants, preferably in insoluble form (immunoadsorbent).

A definition of cross-reactivity can also be assigned to antigens. A cross-reactive antigen is one that binds antibodies induced in response to a different molecule by virtue of shared antigen determinants. Such considerations add nothing further to the concept of cross-reactivity.

## 6. PROPERTIES OF POLYCLONAL ANTISERA AND MONOCLONAL ANTIBODIES

In exploiting the specific binding properties of antibody one has a choice between two very different forms of reagent—these are conventional polyclonal antisera (or often the immunoglobulin fraction) and monoclonal antibodies (Mabs). These forms of reagent are prepared in a different way to satisfy different demands; the choice between them is often not arbitrary but depends to a large degree upon matching the special properties of the antibodies to the assay requirement.

### 6.1 **Polyclonal antisera**

A polyclonal antiserum is the conventional serum product of an immunized animal—usually a rabbit, sheep or goat. It contains many different antibody specificities to the various epitopes of the structurally complex immunogen. The major advantage of polyclonal antisera lies in their capacity to form large insoluble immune complexes with antigen, or to agglutinate cells readily so that the reactions can be seen and measured visually or determined photometrically by, for instance, nephelometry. For all their value, animal antisera have certain limitations for exploitation in immunoassays, the main one being their heterogeneity in specificity, even when reacting to small antigens and their variability between animals and batches. An antiserum is the product of many responding clones of cells and is, in consequence, heterogeneous at many levels; in the classes and subclasses (isotypes) of the antibody produced, their specificity, titre and affinity. In one antiserum there may be antibodies to many discrete antigens (multispecific or polyspecific), to a few antigens (oligospecific) or to a single antigen (unispecific), but even in the latter case the reagent is not homogeneous as single immunogens are still multi-determinant structures which stimulate a polyclonal, multi-determinant-specific response often in several isotypes. In addition the response to individual epitopes may be clonally diverse and antibodies of different affinity may compete for the same epitope. The significance of such heterogeneities within and between polyclonal antisera in practice means that each product is unique in specific antibody composition, in optimal binding conditions and in performance and requires to be separately assessed for its suitability in any particular immunoassay. Without affinity purification specific antibodies in antisera, or in the immunoglobulin fraction, are represented by, at most, 20–30% of the immunoglobulin in most cases. This reduces their efficiency in some procedures and may lead to high background readings in others.

Because of their polyclonal, multispecific nature, conventional antisera cannot be prepared easily or routinely to the degree of specificity needed to determine fine structural and antigenic differences between molecules at the individual epitope level.

### 6.2 **Monoclonal antibodies**

In 1975 Köhler and Milstein (1) developed an alternative method for antibody production which yields monoclonal, monospecific products. These are derived from single antibody-producing cells immortalized by fusion to a B lymphocyte tumour cell line to form 'hybridoma' clones. The secreted antibody is homogeneous in specificity, affinity and isotype and each monoclonal product is specific to a single antigen determinant of the immunogen (monospecific). When prepared *in vitro*, Mabs are the exclusive or predominant protein in the culture medium. Each Mab specificity can be prepared in theoretically unlimited amounts so in appropriate tests universal standard reagents can now be utilized. The application of these antibodies is rapidly extending as more and more laboratories come to recognize their potential and to master the production methods. Already a range of specificities is commercially available, although catalogues so far mainly reflect interest within immunology and microbiology. In these areas the contribution of monoclonal reagents as probes is already very significant. Their special value lies in their ability to select a unique feature of an antigen that may, for instance, allow definition and separation of cell populations and the analysis of bacterial antigens. For assays for soluble antigen, Mabs can undoubtedly solve many of the problems of sensitivity and specificity that accompany the use of polyclonal reagents; they can bring into simple format many assays that were previously complex to perform and they can extend the range of molecules available for investigation. The only major disadvantage of Mabs is that, due to their monospecificity, they lack or have only poor antigen-precipitation properties as single reagents. In consequence assays for antigen in solution using monoclonal reagents are restricted to those that depend upon indirect measurement of antibody–antigen binding, which is the format of the more sensitive assays. The demand for multispecific precipitating antisera and expertise in their production will be untouched by the revolution of Mab technology, as will the importance of the precipitation techniques themselves.

## 7. SELECTION OF ANTIBODY METHODS

It is useful to summarize (*Table 1*) the nature and properties of methods used to pursue the application of antibodies from antibody production and assessment to the final antigen test.

**Table 1.** Properties of antibody methods.

| | *Principle* | *Suitable antibody* | *Sensitivity* [a] | *Advantages/disadvantages* |
|---|---|---|---|---|
| *Precipitation methods* | | | | |
| GDD (Gel double diffusion) | Diffusion of reactants to form visible precipitates in gel | Polyclonal, precipitating | ~10 μg/ml antigen<br>5–20 μg/ml antibody | Informative on antigen complexity and relationships and specificity of antibody. Simple and needs minimum equipment. Low sensitivity and slow result. |
| IEP (Immunoelectrophoresis) | Electrophoretic separation of antigen, and diffusion to form precipitation arcs in gel | Polyclonal, precipitating | 50–100 μg/ml antigen<br>20–100 μg/ml antibody | As for GDD but requires electrophoresis unit. |
| 2D-IEP (Two-dimensional IEP) | Electrophoretic separation of antigens in 2 dimensions, second into antibody-gel to form precipitin peaks | Polyclonal, precipitating —preferably Ig fraction; sheep antibodies very good | Not major consideration and variable according to antigen–antibody system | Multiple estimation of antigens, and antigen-relationships defined. Restricted to antigens of anodic migration. Partially quantitative. Requires cooling plate for gel. |
| RIEP (Rocket IEP) | Electrophoretic migration of antigen into antibody-gel to form precipitation 'rockets' | Polyclonal, unispecific, precipitating | < 5 μg/ml antigen | Rapid estimation of antigen in moderate sensitivity range. Restricted to antigens with anodic migration, and requires unispecific antibody. Requires antigen standards. |
| RID (Radial immunodiffusion in gel) | Diffusion of antigen into antibody-gel to form precipitation rings | Polyclonal, unispecific, precipitating | 1–20 μg/ml antigen | Very simple with minimum equipment. Slow result, more so with high mol. wt antigens. Antigen standards required. |
| RRID (Reverse RID) | Diffusion of antibody into antigen gel to form precipitin rings | Polyclonal, unispecific, precipitating | 1–20 μg/ml antibody | As RID. Slow result, more so with IgM antibodies. Requires reference antiserum. |

**Table 1.** Cont'd

| | *Principle* | *Suitable antibody* | *Sensitivity* [a] | *Advantages/disadvantages* |
|---|---|---|---|---|
| *Agglutination methods* | | | | |
| PHA (Passive haem-agglutination) | Antibody agglutination of antigen-coupled cells performed in microtitre plates | Polyclonal and some monoclonals | 1 – 10 ng/ml antibody | Requires antigen-coupling of red cells and careful standardization. Visual end point. Sensitive and versatile. Tests specificity and titre of antibodies. Applicable to antiglobulin testing with antibody-bound cells or Ig isotype-coupled cells. |
| HAI (Haemagglutination inhibition) | Inhibition by free antigen of antibody agglutination of antigen-coupled red cells | Polyclonal and some monoclonals | 1 – 10 ng/ml antigen | Vies with ELISA and IRMA as antigen assay in mid-sensitivity range. Result less accurate as end point subjective. Can be used as antigen cross-reactivity test. |
| RPHA (Reverse PHA) and Cell rosette assay | Antigen-mediated agglutination of antibody-coupled red cells. Can be used as rosette assay for cell surface antigens | Monoclonal or affinity-purified unispecific polyclonal | 1 – 10 ng/ml antigen | As for HAI. May be problems with prozone. Whole cells can be used as antigen source (for surface marker test) when rosettes are formed with antigen-positive cells. |
| *Enzyme-labelled reagent methods* | | | | |
| EIA (Enzyme immunoassay) | | | | |
| Substrates with chromogenic, fluorogenic or luminescent properties. Competitive assays for antigen | Enzyme-labelled antigen competes with free antigen for solid phase antibody in plastic microtitre plates or on other surfaces e.g. as 'dot blots', beads, pins etc. | Polyclonal, unispecific (as Ig fraction or affinity-purified) or monoclonal, as coating antibody | < 10 ng/ml antigen | Simple competitive principle; can be applied to variety of solid phases. Requires standard conjugated antigen. Can be read visually or to optical density titration end point (requires OD reader). |

| | | | | |
|---|---|---|---|---|
| EMIT (Enzyme-modulated immunotest) | Enzyme-labelled antigen on which enzyme–substrate reaction is sterically blocked by antibody binding. This is inhibited by presence of free antigen. Fluid phase homogeneous test | Polyclonal or monoclonal | μmol–pmol/l for drugs | Usually restricted to measurement of small molecules such as drugs where antigen determinant(s) are close to enzyme coupling site. Rapid sensitive system without washing or separating steps. |
| ELISA (Enzyme-linked immunosorbent assay) Use of labelled antibody to measure antigen or antibody. Solid phases and substrate as for EIA | Enzyme-labelled antibody inhibition assay: Antibody binding to solid phase antigen inhibited by presence of free antigen | Polyclonal or monoclonal purified and labelled | < 10 ng/ml antigen | Simple inhibition principle with standard antibody conjugate. Sensitive quantitative method for antigen. Suitable for 'dot blot' assays. |
| | Sandwich (two-site) assay: Solid phase (antigen-capture) antibody binds antigen and second (enzyme-conjugated) antibody detects bound antigen | Polyclonal or monoclonal or combination, for two antibody layers | < 10 ng/ml antigen | Dual monoclonal system may require different specificities. Potentially very sensitive as can theoretically detect single bound antigen molecule. |
| | Enzyme-labelled antiglobulin assay: Antibody binding to solid phase antigen detected by labelled antiglobulin (called indirect ELISA) | Polyclonal, labelled antiglobulin | < 10 ng/ml antibody | Versatile for titrating antibodies, specificity testing and isotype studies of antibody response. Routine screening system for Mabs. |
| *Radiolabel immunoassays* | | | | |
| RIA (Radioimmunoassay) Competitive assay for antigen in fluid phase | Labelled antigen competes with test antigen for complexing with limited antibody in fluid phase. Precipitate separated prior to counting of precipitate or supernatant | Polyclonal (unispecific) or monoclonal, both with high affinity requirement | < 10 fg/ml antigen (nmol–pmol/l for hormones) | Very sensitive assay requiring stringent specificity of antibody and high affinity. Separation step needed for complexes and expensive equipment. Standard antigen required for labelling. Radiation hazard. |

**Table 1.** Cont'd

| | *Principle* | *Suitable antibody* | *Sensitivity* [a] | *Advantages/disadvantages* |
|---|---|---|---|---|
| IRMA *(Immunoradiometric assays)* | | | | |
| Use of labelled antibody to measure antigens or antibodies | Radiolabelled antibody inhibition assay: | | | |
| | Antibody binding to solid phase antigen inhibited by presence of free antigen. Performed in plastic tubes or flexi-plates with wells separately counted. | Polyclonal or monoclonal as labelled antibody | < 10 ng/ml antigen can be in pg range | Simpler than RIA and washing step replaces separation of complex. Requires radiolabelling of antibody Ig fraction. Standard antigen for inhibition. |
| | Sandwich (two-site) assay: | | | |
| | Solid phase (antigen-capture) antibody binds antigen and second (labelled) antibody detects bound antigen | Polyclonal or monoclonal, or combination, for two antibody layers | < 10 pg/ml antigen | Simpler than RIA and in same sensitivity range. Dual monoclonal system may require different specificities. |
| | Radiolabelled antiglobulin assay: | | | |
| | Antibody binding to solid phase antigen detected by labelled antiglobulin | Polyclonal radiolabelled antiglobulin | < 10 ng/ml antibody | Tests polyclonal or monoclonal first antibody. Routine screening system for monoclonals, for titrating antibodies and isotype studies for antibody response. |

| | | | | |
|---|---|---|---|---|
| WB *(Western blotting)* Identity and purity of antigen: specificity of antibodies | Antigens separated by molecular size by polyacrylamide gel electrophoresis, transferred to nitrocellulose and antigen bands stained with labelled antibody | Monoclonal or polyclonal enzyme- or radiolabelled | | Method reveals purity, molecular size and nature of antigen. Can be used to define and compare specificity of Mabs. |
| *Immunohistochemical and immunocyto-chemical assays* | For antigen detection: Use of fluorescent or enzyme labelled antibodies—often as second antiglobulin reagent—for cell and tissue staining. Enzyme method uses insoluble coloured substrate | Range of discrete antigens detectable greatly enhanced by use of monoclonal (monospecific) first layer antibodies | | Enzyme–substrate systems allow permanent slide preparations and conventional microscopy. Alternative colours available for double antigen marker studies. Immunofluorescence allows detailed cell structure staining as well as double marker (colour) studies. |
| FACS (Fluorescence-activated cell sorting) | For cell population studies and separation | Mainly monoclonal, as fluorescent conjugate or as first antibody with fluorescent polyclonal antiglobulin as second layer | | FACS allows automated fluorescence enumeration of antibody-binding cells with analysis of cell size and marker density. Newer models allow multiple colour (marker) studies. |

[a]Sensitivity is here defined as the minimum detectable amount of antigen or antibody in assays. The data given are approximate and vary according to antigen.

**Table 2.** Choice of antibody method.

| *Requirement* | *Methods* |
|---|---|
| *Immunogen purity* | Gel double diffusion (GDD) |
| | Immunoelectrophoresis (IEP) |
| | Two-dimensional immunoelectrophoresis (2D-IEP) |
| | Polyacrylamide gel electrophoresis (PAGE) |
| *Quality control of antibody* | |
| (a) Specificity | GDD, IEP, 2D-IEP |
| | Immunofluorescence test (IFT) } 1) on standard cells/tissue |
| | Immunoperoxidase test (IPT) } 2) as competitive test versus known antibody |
| | Complement-dependent cytotoxicity (cells) |
| | Passive haemagglutination (PHA) |
| | Enzyme-linked immunosorbent assay (ELISA) |
| | Immunoradiometric assay (IRMA) |
| | Western blotting (WB) |
| (b) Titration | Reverse radial immunodiffusion (RRID) |
| (i) Soluble antigens | PHA (includes antiglobulin test) |
| | ELISA |
| | IRMA |
| (ii) Cells and tissue antigens | IFT |
| | IPT |
| (c) Purity of immunoglobulin fraction of antiserum | IEP |
| (d) Specificity/titre of antiglobulin reagent | ELISA, IRMA, PHA |
| (e) Isotypes of antibody | GDD |
| | ELISA } |
| | IRMA } labelled anti-isotype |
| | IFT } |
| | IPT } anti-isotype staining |

| | |
|---|---|
| *Antigen assay methods* | |
| (a) Antigenic relationship between molecules in solution | GDD, IEP, 2D-IEP |
| | IRMA and RIA |
| | Enzyme immunoassay (EIA) |
| | Haemagglutination inhibition (HAI) |
| (b) Complexity of antigen mixtures in solution | GDD, IEP, 2D-IEP, WB |
| (c) Quantitation of antigen mixtures in solution | 2D-IEP |
| (d) Identity and purity of antigens in low concentration | Western blotting of polyacrylamide gel |
| (e) Quantitation of single antigens in solution | Radial immunodiffusion (RID) |
| | Rocket immunoelectrophoresis (RIEP) |
| | HAI |
| | Reverse passive haemagglutination (RPHA) |
| | EIA and ELISA |
| | RIA and IRMA |
| (f) Detection of antigen on cells and tissues | IFT, IPT |
| (i) Cell suspension population analysis | IFT, IPT |
| | Fluorescence-activated cell sorting (FACS) |
| | Reverse passive rosetting with antibody-coupled red cells |
| (ii) Tissue typing: red cells | Direct haemagglutination (see Volume II) |
| lymphocytes | Complement plus antibody-mediated cytotoxicity (see Volume II) |
| | IFT |
| | Other new methods (see Volume II) |

Working with antibodies often requires self-produced reagents and for this the first task is the preparation of an immunogen. For soluble single immunogens the molecules must be as pure as possible to achieve good results and the first antibody method will be a test for purity. After harvesting antiserum the quality of the reagent should be assessed, in relation to the final test system, for specificity, titre and binding characteristics. This demands a knowledge of quality control procedures including in some cases determination of the constituent antibody isotype(s). For the final application of the antibody a choice of test methods will be available (*Table 2*)—for instance between immunofluorescence and immunoperoxidase for tissue and cell antigens, or between agglutination, enzyme or radioimmunoassays (RIAs) or precipitation tests for soluble antigen. For some test systems the antibody may need to be purified for labelling or an antiglobulin reagent prepared and labelled. The advantages and disadvantages of test systems require careful consideration.

## 8. REFERENCES

1. Köhler,G. and Milstein,C. (1975) *Nature*, **256**, 495.

CHAPTER 2

# Production and quality control of polyclonal antibodies

DAVID CATTY and CHANDRA RAYKUNDALIA

## 1. INTRODUCTION

As different tests require different properties of antisera there can be no single procedure for animal immunization that guarantees an ideal product for all requirements. Nevertheless certain principles can be adopted which together form the 'ground rules' for antiserum production; ideal reagents rely to some degree on trial and error as every immunogen is different and every animal responds individually. The essential properties of antisera are high titre coupled with high average avidity and specificity. The former, which are influenced by the use of various adjuvants, depend upon achieving a subtle balance between widespread stimulation of antigen-sensitive cells of the animal's lymphoid tissues, continued presence of immunogen to allow extensive polyclonal proliferation of plasma cells but, within this, a competition between cells for limiting amounts of antigen so that only the highest affinity cells will respond. The latter depends critically upon the purity of immunogen administered, as minor contaminants may induce a disproportionately large antibody response.

Many antisera are produced for their precipitating properties and since this depends upon efficient lattice formation between antibodies and several antigen determinants, both multiple specificity and a balance in the titre of antibodies is important. Antisera that can be used in high dilution are always preferable to reduce background activity. In 'rocket' and two-dimensional immunoelectrophoresis (RIEP and 2D-IEP respectively) it is essential to have antibodies in the agar gel of narrow isoelectric range and the choice of immunized species is important here—sheep $IgG_1$ antibodies have especially good qualities in this respect.

Where antisera are to be used in more sensitive quantitative assays such as enzyme-linked immunosorbent assay (ELISA) or radioimmunoassay (RIA) there are more stringent requirements for high avidity and specificity. Antiglobulin reagents must bind only to the target species immunoglobulin, or to a particular isotype of that species if required, and scrupulous absorption of species and isotype cross-reactive antibody components will be needed to achieve this.

Most applications of RIA are for measurement of small molecules at low concentration. For immunization, small molecules, having low intrinsic immunogenicity, require coupling to larger protein carriers. Methods for such coupling are important considerations in determining the specificity of the response. Careful selection from several antisera may be needed to find one that has suitable properties.

The harvesting of an antiserum is only the start of obtaining a useful working reagent.

For most assays the only relevant part of the serum is what lies in the immunoglobulin fraction—the rest can be conveniently removed. This increases the ratio of antibody to total protein, reduces background reactions and increases sensitivity of assays. Immunoglobulin separation is also the initial step for fluorochrome, enzyme and radioisotopic labelling of antibody, and in affinity purification.

After initial tests in specificity and titre, quality control steps may involve absorption of unwanted antibodies as an alternative to affinity purification. This is best performed with insolubilized antigen polymers or beads, as fluid absorption forms soluble immune complexes with unfortunate properties.

Once the specificity and titre of the antibody is satisfactory it can be calibrated against a reference reagent and finally tested across a range of immunoassays in which it might be applied. *Table 1* summarizes the strategy for antibody production and quality control.

## 2. GROUND RULES IN ANTIBODY PRODUCTION

### 2.1 **The immunogen**

#### 2.1.1 *Small molecules*

Molecules of less than 5 – 10 000 mol. wt require conjugation to a carrier protein. The antibody response will be directed towards the carrier determinants as well so choose a carrier that is irrelevant to future assays and/or that can be prepared easily as an adsorbent. Although use of a protein carrier from the immunized species might seem a useful tactic this is inadvisable as antibody responses to most (thymus-dependent) immunogens require recognition of carrier determinants by helper T cells. Bear in mind that the size, charge and polar or non-polar properties of haptens are important in inducing the antibody responses and these properties can be influenced by the extent of uniformity and density of conjugation of hapten to carrier and by the choice of carrier. It may be necessary to prepare the conjugates with several carriers and with a range of hapten-carrier coupling ratios and to try all these to find the best constructed immunogen. In principle, antibodies to larger haptens may be hapten-specific but these are usually also heterogeneous. The choice of coupling reaction and to which group of the hapten (see Section 3.1), is critical to hapten orientation, and will influence antibody specificity, which is usually away from the coupling site. Antibodies to smaller haptens may incorporate specificity to the link region. It may be necessary to conjugate using a 'linker' group and in any event the anti-hapten specificity should not be tested with the same conjugate but with free hapten [e.g. in RIA or haemagglutinin inhibition (HAI)] or hapten conjugated in the same way to a different carrier.

#### 2.1.2 *Large molecules*

Those with a higher degree of conformation and structural rigidity tend to be stronger immunogens with immunodominant regions and multiple antigenic determinants which induce precipitating antiserum. The use of adjuvant is nevertheless recommended for these molecules to obtain the best results.

**Table 1.** Strategy for antibody production and quality control.

| | | |
|---|---|---|
| 1. | *Preparation of immunogen* (Sections 3 and 10) | |
| | (a) | Purify. |
| | (b) | Test purity. |
| | (c) | Conjugate to carrier if less than 5000 to 10 000 mol. wt. |
| 2. | *Immunization* (Sections 2, 4 and 6) | |
| | (a) | Select an appropriate species as recipient. |
| | (b) | Decide on the form of immunogen presentation (e.g. with adjuvant), route of administration, dose and timing; allow as long as possible between injections – up to several months (advance planning is essential). |
| | (c) | Prepare the immunogen in final format for injection (emulsion, precipitate etc.), enough for a single dose. |
| 3. | *Test procedure* (Sections 5, 8 and 9) | |
| | (a) | Test bleed and separate the serum. |
| | (b) | Check specificity, titre and binding properties; include tests under same conditions as planned for final application. |
| | (c) | Do trial absorptions on basis of any detected cross-reactivities—prepare insoluble adsorbents. |
| | (d) | Retest absorbed serum. If satisfactory prepare for scale-up absorption. |
| | (e) | Decide on boosting strategy on basis of interim tests. |
| | (f) | Repeat (a) to (d) as required. |
| 4. | *Harvesting, storage and absorption* (Sections 5, 7 and 9) | |
| | (a) | The volume required determines staged harvesting or immediate sacrifice of animals. |
| | (b) | Store in aliquots labelled according to batch/date—store at −20°C or preferably lower. |
| | (c) | Test- and trial-absorb subsamples, and then bulk-absorb aliquots as required. |
| 5. | *Immunoglobulin fractionation* (Section 10) | |
| | (a) | Prepare by the two-stage method (salt precipitation and anion-exchange chromatography). |
| | (b) | Test the purity of the fraction. |
| | (c) | Check the activity of the fraction. |
| | (d) | Conditions for storage vary by species. |
| 6. | *Affinity purification of antibodies (when required)* (Section 10 – see also Chapter 5) | |
| | (a) | Check the purity of the antigen. |
| | (b) | Prepare the antigen affinity column by appropriate coupling chemistry to selected matrix. |
| | (c) | Check antigenic activity of the column, loading capacity and conditions for antibody elution. |
| | (d) | Perform trial purification, testing purity, activity and stability of eluted product. |
| | (e) | Perform bulk purification with appropriate eluate storage in aliquots. |
| 7. | *Quality control* (Section 11) | |
| | (a) | Assess the performance of the product across a range of applications—use antigen standards where available. |
| | (b) | Calibrate against a reference reagent where applicable. |
| | (c) | Quantitate antibody concentration (to soluble single antigens). |
| | (d) | Determine isotype(s) of antibody. |

### 2.1.3 *Foreignness*

There is an overriding necessity for a molecule to be seen as foreign by the recipient animal. Mammalian serum proteins, for instance, although in the main structurally complex, are poor immunogens in closely-related species, requiring assistance of both adjuvant and T helper cells to stimulate antibody production (T-dependent antigens).

**Table 2.** Major properties of adjuvants.

| | |
|---|---|
| 1. | They potentiate antibody production by increasing the efficiency of antigen presentation and the number of collaborating and secreting cells involved, effectively reducing the optimum required immunogen dose. |
| 2. | They enhance immunogenicity, allowing antibody responses to molecules with borderline immunogenic properties. |
| 3. | They alter the isotype pattern of antibody responses. |
| 4. | They prolong antibody responses by an immunogen depot effect which protects the immunogen from rapid removal and breakdown, thereby reducing the need for repeated injections. |
| 5. | They increase the average avidity and affinity of the antibody response. |

The response to species-related proteins is in antibodies that recognize only the minor structural differences that exist between the same molecule in the two species. This may be usefully exploited in some circumstances, even to the extent of raising antibodies to allelic structural (antigenic) differences between molecules within the same species. By contrast the broadest set of antibody specificities to a protein is produced in species phylogenetically distant to the immunogen source. This is particularly relevant in raising antibodies to complex thymus-dependent soluble or cell-bound antigens of mammalian origin. Both sheep and rabbits respond well to human and other animal proteins, glycoproteins and lipoproteins and to antigens of human and other animal pathogens. An important point to bear in mind is that immunization with microorganisms or their extracted antigens may induce an increase of an existing antibody response to antigens of the recipient's own microbial flora by virtue of shared (cross-reactive) epitopes.

## 2.2 **Use of adjuvants**

Adjuvants (L. *adjuvare*—to help) consist of a number of substances whose physical and/or biological properties when applied with immunogen affect to different degrees the antibody response in one or more of the ways shown in *Table 2*. The properties of adjuvants have been extensively reviewed (1–3). Adjuvants find their major uses in preparing antisera to soluble thymus-dependent immunogens where by their application, optimal immunizing doses may be reduced to microgram quantities. This complies with the rule, determined from many observations, that the smallest practicable dose of immunogen yields, by patient long-term application, the best quality antiserum in terms of titre and affinity. The following are the most popular types of adjuvant for animal use.

### 2.2.1 *Freund's Complete Adjuvant (FCA or CFA)*

This consists of a mixture of mineral oil (Bayol F), a suspension of heat killed *Mycobacterium butyricum* or *Mycobacterium tuberculosis* and the emulsifier Arlacel A. Immunogens in solution are emulsified with these components to produce an antigen/water microdroplet phase within the oil phase. FCA greatly enhances and prolongs the antibody response as first described by Freund (4). The emulsion is stable and when injected intramuscularly forms a depot of immunogen which only slowly becomes available. The other major attribute is related to the activity of the mycobacteria, some of whose products act as powerful stimulants for cells of the immune system both at local lymph nodes and in the granuloma that forms around the depot. *M.tuberculosis* is the more potent of the two mycobacterial species used. Much antibody is synthesized within the granuloma and this activity continues so long as immunogen persists, which may be for many years. Because of granuloma formation and the tendency for FCA to render the body's own tissues autoimmunogenic it cannot be used in man. It is, however, by far the most effective and widely used adjuvant for routine antiserum production in animals to soluble antigens and antigens of emulsified cells. FCA induces an almost exclusive IgG antibody response in the rabbit, a predominant $IgG_1$ subclass response in sheep and $IgG_2$ antibodies in the guinea pig. FCA and Freund's Incomplete Adjuvant (FIA) (which lacks mycobacteria—see below) can be purchased from Difco Ltd (see Appendix). Their preparation from components, emulsification and injection methods are given in Section 4.1.

### 2.2.2 *Freund's Incomplete Adjuvant (FIA or IFA)*

Water phase immunogen can be emulsified with mineral oil without mycobacteria to provide a depot immunization principle. This also provides an intense long term antibody response as there is only gradual uptake and loss of immunogen from the injection site. However, the response with FIA is commonly found to be less reliable and of lower order—this may be due to the fact that open active granulomas do not develop and persist at the FIA depot site. There may also be differences in the antibody isotype response using FIA exclusively. In guinea pigs the response to soluble proteins is in $IgG_1$. FIA is commonly used for booster immunizations in subcutaneous sites because of its reduced pathogenic effects.

### 2.2.3 *Alum-precipitated immunogen*

The use of potassium aluminium sulphate to precipitate proteins at alkaline pH (5) is a useful method for extending immunogen availability and prolonging the antibody response. When used intramuscularly it is claimed to induce plasma cell accumulation at the depot site. Its side effects, however, are minimal and it is used in man for diphtheria and tetanus toxoid vaccination. Immunogen precipitates can be injected subcutaneously or by the intraperitoneal route as well as into muscles. Its favoured use is in mice, often mixed with a *Bordetella pertussis* suspension (see Section 2.2.4) and one of its major advantages is that it allows fairly rapid dissemination of immunogen without the risks attached to injection of molecules in solution into primed animals which can lead to rapid fatal anaphylaxis. A commonly used strategy is to divide alum precipitates into very small doses of a few μg and give these every few days over several weeks. Alum-precipitated toxoids and other immunogens are known to induce IgE antibodies in some species. The preparation of alum protein precipitates for immunization is described in Section 4.2.

### 2.2.4 *Bordetella pertussis*

A suspension of killed *B.pertussis* injected jointly with immunogen is another good method of obtaining a potentiated antibody response, particularly in mice. The principle active components of pertussis are the lipopolysaccharide (LPS) and pertussis toxin.

The toxin is known to enhance both IgG and IgE antibody production (6).
Killed *B.pertussis* suspensions can be purchased commercially (see Appendix).

### 2.2.5 *Liposomes*

Artificial liposomes from lecithin, cholesterol and stearylamine can be prepared (7). The antibody response to diphtheria toxoid incorporated within these is enhanced. Use of liposome-entrapped immunogen allows a rapid localization of molecules by the spleen after intravenous injection or a prolonged release from an intramuscular subcutaneous or intradermal depot. In the former case adjuvant activity may be related to an efficient presentation of packaged immunogen on the liposome membrane to antigen-sensitive cells. Adjuvant activity can be increased by the addition of LPS, lipid A or muramyl dipeptide (Section 2.2.6) to the liposome contents in addition to the immunogen. One important application of liposomes is immunization with acutely toxic substances such as snake venoms. High, sustained antibody responses to otherwise toxic doses of carpet viper venom in mice, rabbits and sheep have been induced by intravenous and subcutaneous injection of venom entrapped in stabilized liposomes (8,9).

### 2.2.6 *Muramyl dipeptide*

The adjuvant active compound of mycobacteria has been identified as a tripeptide-monosaccharide, from which the smallest active component has been defined as muramyl dipeptide (MDP) (10). This is now obtainable commercially (see Appendix), and can be used to replace heat-killed mycobacterium species in FCA. It is also a potent adjuvant for the antibody response when administered with immunogen in saline when it induces an exclusive $IgG_1$ antibody response in mice. It also promotes antibody responses to synthetic antigens.

## 2.3 **Choice of animal**

### 2.3.1 *Mice and rats*

Much of our knowledge of the nature of the immune response derives from studies in the laboratory mouse and rat. These animals are the best suited to intensive husbandry, they are available in many inbred genetic strains of relevance to immunologists, but also show uniform response characteristics and so, in batches, can be reliable sources of antibody. Their other special contribution is that, through inbreeding, allelic differences in many naturally polymorphic molecules are fixed within the strains so they offer unique opportunities for studying allelic systems such as the histocompatibility gene complex, allotypes and isoenzymes; cross-immunization between strains induces allele-restricted antibody responses. In addition the mouse and rat are the only two species of laboratory animal that can be utilized for *in vitro* cell fusion to yield Mab-secreting hybridomas. The mouse and rat therefore occupy a special place in antibody production, and methods for their immunization and handling require consideration. Inbred rats and mice show strain differences in response to molecules of low immunogenicity so choice of strain may be very important.

### 2.3.2 *Rabbits*

Production of polyclonal antisera for routine immunoassay work, especially in precipitation methods, relies upon larger animals and in particular the laboratory rabbit. Rabbits are amenable to cage husbandry, breed well in captivity, are robust, long-lived and easy to handle, immunize and bleed. They make excellent precipitating antisera which are very stable on storage. The immunoglobulin fraction is easy to purify. With patience rabbits produce antibodies with high avidity/affinity characteristics for sensitive antigen assays. At least 100 ml of antisera can be harvested from a rabbit over several months at peak responses with a final blood exsanguination volume of up to a further 150 ml. Because they are an outbred species rabbits respond very individually, so more than one animal should be used for each immunogen, their sera tested individually and selected samples pooled where maximum diversity is required. Rabbits give excellent IgG responses to a wide variety of immunogens with long-term Freund's adjuvant immunization.

In certain circumstances, such as chronic intravenous immunization with killed pneumococcal suspensions, they can produce IgG anti-polysaccharide responses in excess of 50 mg/ml of specific antibody, but 1−2 mg/ml antibody to protein antigens is a usual high titre response. This is, however, far in excess of that obtained in the mouse where IgG antibodies have a half life of only 1−2 days, are much less stable, especially in purified form and precipitate poorly with antigen. Mice yield only a few ml of blood and require sacrifice to obtain this.

#### 2.3.3 *Sheep*

For those with access to sheep this species is to be thoroughly recommended for bulk production of antisera in excess of 5 litres. Sheep antibodies, like those of the rabbit, have excellent precipitating properties. With FCA, the response is predominantly $IgG_1$ which is stable, stores well, is easy to purify and has little or no mobility at pH 8.6 in agarose gels for RIEP and 2D-IEP. On a volume basis sheep antibodies are the cheapest to produce. Sheep are easy to immunize and bleed and with plasma expanders and blood cell replacement routines there is no requirement for sacrifice. Young breeding stock can be immunized over the Spring and Summer of their first season and bled in the Autumn and Winter without prejudice to their future breeding performance.

For antiglobulin production both the rabbit and sheep make excellent responses to human immunoglobulins of all classes. Precipitating anti-human IgG subclass reagents have been made successfully in sheep which illustrate the fine discriminating properties of this species for small structural differences in other mammalian immunoglobulins. It is convenient that both sheep and rabbits make similarly good antisera to rat and mouse isotypes and in addition, being phylogenetically distant, they make excellent antiglobulins to each other's molecules. Thus an assay using a primary rabbit antibody can be developed with a labelled sheep anti-rabbit immunoglobulin and vice versa. Affinity-purified rabbit and sheep antibodies are stable on storage. In addition Fab and $F(ab')_2$ subunits of antibodies can be easily and reproducibly prepared from antibodies of these two species by well-described techniques (11).

### 2.4 Guidelines on the use of animals

#### 2.4.1 *Numbers*

Always use a number of animals for immunization with any immunogen and test each response individually. This takes account of variation between animals and allows pooling to provide maximum diversity where required or selection of the highest avidity/affinity response.

#### 2.4.2 *Condition*

Never use animals in poor condition or those under stress. For laboratory animals ensure that the group is maintained under proper environmental conditions with a minimum of noise.

#### 2.4.3 *Age and sex*

As a general rule use young adult animals and avoid aged stock. Mice and rats of 3–4 months of age and rabbits of 4–6 months are ideal. There is no characteristic difference in antibody response between the sexes but never immunize pregnant animals, unless this forms part of a licensed experiment where pregnancy is part of the aim.

#### 2.4.4 *Legal responsibilities*

Home Office regulations in the UK demand that *all procedures* related to immunization and subsequent bleeding must be covered by a personal licence and an approved project licence. The scientist holds personal responsibility for the welfare of the animals which must not be delegated. Proper records of all procedures must be maintained. Methods for immunization, anaesthesia, bleeding and sacrifice are carefully prescribed to minimize pain and stress. Use of sheep also requires licensing. No work on preparing antisera in animals (or Mabs) can proceed without appropriate licences. Licensees are required to provide evidence of appropriate training in animal methods. Similar strict regulations on the use of animals for antibody production exist in all member countries of the European Economic Community and in the USA.

## 3. PREPARATION OF IMMUNOGENS

### 3.1 **Immunogens prepared by hapten–protein conjugation**

The production of antibodies to low molecular weight molecules requires their chemical attachment to an immunogenic carrier which is usually a protein. The use of a complex protein carrier has the advantage that, being thymus-dependent, the compound immunogen involves the induction and proliferation of T helper cells which collaborate in producing the anti-hapten response. About 10–30 haptenic groups per 100 kd of carrier is desirable to achieve a good antibody response.

Some commercially available haptens couple spontaneously to the amino groups of proteins and because of their ease of preparation and antigenic simplicity they have a special place in studies of antibody specificity and the cellular basis and kinetics of antibody production. They are ideal models for teaching purposes. Examples of commonly used haptens in this category are the dinitrophenyl (DNP) and trinitrophenyl (TNP) groups. Commonly used carriers are heterologous gamma globulins, albumins and thyroglobulins; keyhole limpet haemocyanin (KLH) and *Maia squinada* haemocyanin (MSH).

For small molecules that do not couple spontaneously to carriers it is necessary to perform a chemical activation or bridging step. Several alternative approaches are available using a number of activating agents; the choice will depend on the functional groups available on the candidate hapten, the required hapten orientation, distance from the carrier and effect of conjugation on biological and antigenic properties. An excellent discussion of this topic is provided by Parker (12). Details of conjugation methods not covered below are provided in two useful reviews (13,14). A few examples are given here to illustrate the major approaches.

3.1.1 *Spontaneous coupling methods*

(i) *TNP–bovine γ-globulin (TNP–BGG).* 2,4,6-Trinitrobenzene sulphonic acid (picryl sulphonic acid, TNBS) reacts with proteins at pH 10–11 at room temperature in aqueous solution, the sulphate being split out and the TNP being substituted in the protein amino groups to form the TNP hapten.

(1) Dissolve 50 mg of BGG in 5 ml of 0.1 M $NaHCO_3/Na_2CO_3$ buffer, pH 9 in a glass 20 ml universal bottle.
(2) Weigh out 25 μg of TNBS/mg of protein carrier (i.e. 1.25 mg) and dissolve in 1 ml of the same buffer. Prepare a short length (10 cm) of narrow visking tubing by soaking in distilled water, knot tightly at the one end, add the hapten solution to the bag by Pasteur pipette and seal the other end to make a small dialysis bag, cutting the unused tubing behind the knots.
(3) Place the bag in the protein solution with a small magnetic bar and stir slowly in an ice bath on a magnetic stirrer overnight at 4°C.
(4) Make a small Sephadex G25 (Pharmacia) column (30 × 1.25 cm) equilibrated with phosphate-buffered saline (PBS), pH 7.2 and pass the mixture through this using PBS. Collect the first visible yellow band of conjugate. Free hapten is retained at the top of the column.
(5) To determine the molar ratio of TNP:BGG in the conjugate read an appropriate dilution (~1:30) in a spectrophotometer at the peak absorption wavelengths (λ max) for the hapten (355 nm) and protein (280 nm).
1 mg/ml TBP (mol. wt 229), 1 cm cell = 62.9.
1 mg/ml BGG (mol. wt 150 000), 1 cm cell = 1.45.

In an example conjugation with the above starting amounts of reagents, the following data were obtained:

OD 355 nm at 1:30 dilution = 0.137
OD 280 nm at 1:30 dilution = 0.185

$$\text{TNP content of conjugates} = \frac{0.137 \times 30}{62.9} = 0.07 \text{ mg/ml}$$

$$\text{BGG content of conjugate} = \frac{0.185 \times 30}{1.45} = 3.8 \text{ mg/ml}$$

$$\text{Molar ratio hapten protein} = \frac{0.07}{229} : \frac{3.8}{150\ 000} = 12$$

This is an underestimate as TNP on the protein accounts for approximately 40% of the 280 nm absorbance. In consequence the ratio is closer to 20.

(ii) *DNP–Maia squinada haemocyanin (DNP–MSH).* A simple method of preparing DNP conjugates is as follows (15).

(1) Take 50 μg of MSH and dissolve it in 10 ml of 0.1 M borate buffer, pH 8.5.
(2) Make this solution 10% (w/w) in an oil-based solution of 2,4-dinitrofluorobenzene (DNFB).
(3) Warm this mixture to 37°C to bring the hapten into aqueous solution. Stir at room temperature for 1 h and then extensively dialyse against 0.9% saline at 4°C to remove the free DNFB.

This gave in the reported work a substitution ratio DNP:MSH of 150, assuming the molecular weight of MSH to be $10^6$.

3.1.2 *Haptenation by activation and bridging reactions*

Three major approaches to haptenation to carriers are available.

(1) Introduction of a bridging agent between amino acid groups of hapten and protein carrier. This can be accomplished with diisocyanates, halonitrobenzenes and imido esters.
(2) The use of bifunctional diazonium salts to bridge tyrosyl, histidyl or lysyl residues.
(3) The activation of carboxyl groups of the hapten to react with amino groups of the protein carrier to produce a CO–NH bond. This involves the use of agents such as the carbodiimides, alkyl chloroformates and isoxazolium salts.

(i) *Conjugation of urease to sheep IgG by the one-step glutaraldehyde method*

(1) Dialyse 5 mg of purified sheep IgG in 2 ml overnight against 0.1 M phosphate buffer, pH 6.8 at 4°C. Add to this with stirring 20 mg of urease enzyme and 50 μl of 1% glutaraldehyde solution.
(2) Hold the mixture at room temperature for 2 h and then add 200 μl of 1 M lysine solution, pH 7, to quench remaining reactive sites. After a further 2 h dialyse the mixture against PBS pH 7 at 4°C overnight and centrifuge at 40 000 *g* for 1 h.
(3) To separate unreacted urease pass the conjugate through a Sephadex G25 column in PBS and collect the first protein peak.

(ii) *Conjugation of aminoglycocide drugs (amikacin, gentamicin) to porcine thyroglobulin (PT) or BSA using water soluble carbodiimide (16)*

(1) Dissolve 40 mg of the drug in 0.5 ml of distilled water and adjust the pH to 5 (if necessary) by addition of 1 M HCl. To this add 20 mg of carrier, allowing it to dissolve slowly without frothing.
(2) Freshly dissolve 100 mg of 1-ethyl-3(3-dimethylaminopropyl) carbodiimide HCl in 0.5 ml of distilled water and add this to the drug/carrier solution and incubate for 2 h at room temperature with occasional mixing.
(3) Dialyse the mixture for 2 days at 4°C against several 1 litre changes of PBS, pH 7.2, to remove unconjugated reagent.

Using tritiated aminoglycocides (Amersham International), diluted with unlabelled hapten in known proportion, it was determined that the above conjugation leads to 20–90 haptenic groups per thyroglobulin molecule (650 000 mol. wt) and 5–20 per albumin molecule (66 000 mol. wt). This was estimated after separating unreacted drug from conjugate by passing the mixture down a Sephadex G25 column in PBS and collecting the first, major, protein peak.

(iii) *Conjugation of β-D-galactosidase to sheep IgG using N,N'-O-phenylenedimaleimide*

(1) Dialyse 14 mg of sheep IgG in 2 ml against 0.1 M sodium acetate buffer pH 5.0 overnight and centrifuge at 3000 r.p.m. for 20 min to remove any insoluble protein. Reduce this by addition of 10 mM (2.27 mg) 2-mercaptoethylamine and incubate at 37°C for 90 min.
(2) Separate the reduced IgG from compounds of lower molecular weight by passing through a 1 × 40 cm Sephadex G25 column equilibrated with sodium acetate buffer.
(3) Determine the amount of IgG by absorbance at 280 nm (1 mg/ml:1 cm = 1.45).
(4) Add 3 mg in 1 ml of reduced IgG dropwise to 1 ml of a saturated solution of *N,N'-O*-phenylenedimaleimide (0.75 mM) at 0°C. Incubate this mixture at 30°C for 20 min and then pass it through a Sephadex G25 column to remove the unreacted coupling agent.
(5) Adjust the treated IgG concentration to give an OD 280:1 cm reading of approximately 1.0. Incubate each 1 ml with 20 μl of β-D-galactosidase at 5 mg/ml for 20 min at 30°C. Neutralize the mixture with 1 M NaOH by adding 15–20 μl/ml.
(6) Add 2 μl each of 5% BSA and 1 M $MgCl_2$ to stabilize the enzyme. Keep the mixture at 4°C for 72 h and purify the conjugate by passing through a Sepharose 6B column (1 × 40 cm) equilibrated with 0.01 M sodium phosphate buffer pH 7 containing 0.1 M NaCl, 1 mM $MgCl_2$, 0.1% sodium azide and 0.01% BSA, using this buffer for elution and storage.

(iv) *Conjugation of glycocholic acid to BSA by the mixed anhydride technique (17)*

This is a general method described for the preparation of steroid–protein conjugates.

(1) As in our application (18), dissolve 537 mg (1.1 mmol) of glycocholic acid in 10 ml of dioxane. Add 0.25 ml (1.1 mmol) of tri-*n*-butylamine followed by 0.14 ml (1.1 mmol) of isobutyl chlorocarbonate. Allow the reaction to proceed for 20 min at 4°C and then add the mixture to 1.5 g (0.02 mmol) of BSA in 85 ml of water:dioxane (1:1 v/v) containing 1.5 ml of M NaCl.
(2) After 1 h, add 2 ml of 1 M NaOH and continue the reaction for a further 3 h at 4°C. Add ice-cold acetone to the reaction mixture and collect the resulting precipitate by centrifugation at 5000 *g* for 30 min at 4°C.
(3) Suspend the precipitate in $NaHCO_3$, pH 8.5 and repeat the treatment with ice-cold acetone. Resuspend the precipitate in $NaHCO_3$, pH 8.5 and dialyse against 0.5 M phosphate buffer for 48 h. The latter step ensures complete removal of free glycocholic acid from conjugate. Analysis of the buffered conjugate solution with 3α-hydroxysteroid dehydrogenase has shown that approximately 12 moles of glycocholic acid is bound per mole albumin.

(v) *Conjugation utilizing haptenic aromatic amino group conversion to a diazonium group*

The reaction occurs in two stages—the diazotization of the amino group of the hapten to diazonium chloride with HCl and $NaNO_2$, which then couples spontaneously to tyrosines of the protein carrier. The use of a prepared bifunctional diazonium salt (bisdiazotized benzidine) allows a spontaneous bridging reaction to occur between haptens with aromatic amino acids (e.g. peptides) and proteins. The two step reaction is exemplified by the conjugation of para-aminobenzoic acid to BSA.

(1) Dissolve 100 mg of BSA in 10 ml of 0.5 M borate buffer pH 9.

(2) Treat 15 mg of para-aminobenzoic acid with 1.6 ml of 0.2 M HCl and make the solution up to 5 ml with distilled water. Cool to 0°C in an ice bath.

(3) Dissolve 8 mg of $NaNO_2$ in 0.5 ml of distilled water at 0°C and add this to the para-aminobenzoic acid solution to allow diazotization. Free nitrous acid is generated which should be tested by the blueing of starch iodide paper. An excess of HCl is required to form the amine hydrochloride and the generation of nitrous acid to allow the subsequent coupling reaction to occur in acid conditions.

(4) Add the benzoic acid paradiazonium chloride dropwise into the BSA solution. After 1 h neutralize the solution and remove any unconjugated diazonium compound by extensive dialysis.

(vi) *Conjugations utilizing haptenic non-aromatic amino groups through a thiolation reaction*

This popular method involves the separate pre-treatment of hapten and protein carrier, coupling their amino groups with *N*-succinimidyl-3(2-pyridyldithio) proprionate (SPDP) at pH 7.5.

We have described the following protocol for conjugation of the enzyme horseradish peroxidase (HRP) to IgG (19).

(1) Dissolve 10 mg of HRP in 2 ml of sodium phosphate buffer 0.1 M pH 7.5, containing 0.1 M NaCl. Add to this 400 $\mu$g of SPDP dissolved in 0.5 ml of absolute ethanol, dropwise with stirring. Allow the mixture to react for 30 min at room temperature with occasional further stirring.

(2) Remove excess SPDP and the reaction product *N*-hydroxysuccinimide from the 2-pyridyl-disulphide-substituted enzyme by gel filtration through a 1.6 $\times$ 35 cm Sephadex G25 column, eluting the substituted HRP in the first protein peak with PBS as eluant.

(3) An identical reaction can be performed with SPDP and IgG, using 10 mg of protein and in this case only 10–15 $\mu$g of SPDP in 0.2 ml of ethanol. The substituted IgG is again eluted from a Sephadex G25 column in PBS.

(4) Generate reactive thiol groups on the substituted enzyme by reduction of the 2-pyridyl disulphide groups with DTT using 2.5 mM/mg substituted enzyme at room temperature. Remove the excess reducing agent and pyridine-2-thione by further gel filtration.

(5) Mix the thiol-containing peroxidase and the 2-pyridyl-disulphide-IgG solutions at a 1:1 (w:w) proportion and leave at 4°C for 18 h.

(6) Separate conjugate and uncoupled treated enzyme by passage through a Sephacryl S-200 column (1.6 × 95 cm) using PBS as eluant. The first protein peak (enzyme active) contains conjugate with a molecular weight of approximately 200 000.

(vii) *Conjugations using carbohydrate haptens or carriers*

There are several methods for conjugation of carbohydrate haptens to carriers which depend upon the use of their reducing ends, the carboxyl groups (of acidic carbohydrates) and hydroxyl groups (14).

Where carboxyl groups are absent, these can be introduced by carboxymethylation using chloracetate which binds to hydroxyl groups to generate carboxyl (20). By this manoeuvre many carbohydrates can then be coupled to proteins via the carboxyl group, for instance by the carbodiimide reaction, in the same way as acidic polysaccharides.

Carboxymethylated and acidic carbohydrates (with internal carboxyl groups) are popular carriers for haptenation in studies of antibody responses to thymus-independent synthetic immunogens. An aminoethylation step further prepares these carriers for conjugation to spontaneously coupling haptens such as TNBS. Aminoethyl-carboxymethyl−Ficoll (AECM−Ficoll) coupled to TNP is a favourite thymus-independent immunogen. The following method may be used for its preparation (14).

(1) Dissolve 40 mg of AECM−Ficoll in 2 ml of 0.1 M $NaHCO_3$. To this add 50 μl of TNBS dropwise at a concentration of 26 g/litre.
(2) Stir the mixture for 18 h at 4°C and then dialyse against saline at 4°C for 4 days to remove uncoupled hapten. This yields about 4.2 molecules of TNP per 100 kd of carrier. AECM−Ficoll has a molecular weight of approximately 400 000.

## 3.2 Preparation of macromolecules for immunization

The range of macromolecules potentially available for study by antibody techniques is enormous and covers not only the naturally occurring cell products of vertebrate and invertebrate animals and of plants, but also microbial products and synthetic compounds. Where antibodies to individual molecules are required the production of polyclonal antisera may demand, when a commercial pure source of immunogen is unavailable, preliminary work in purification. For cell surface molecules the contemporary approach is through production of Mabs and selection of the required specificity which resolves the problem of immunogen purification (see Chapter 3). Alternatively, methods are available to purify cell-surface components in situations where an antibody already exists and to enrich for cells expressing the target antigen. These approaches are discussed in Chapter 5. Where multispecific antisera are to be raised against a mixture of immunogens, for instance to extracted cell membranes, tissue homogenates, bacterial, viral or parasite sonicates, the method for extraction may be critical. It is not possible in this chapter to provide detailed guidance on extraction and purification methods for the very wide range of possible immunogens. Excellent specialist chapters on preparation of carbohydrate, bacterial, viral, fungal and parasite antigens have been extensively reviewed (21). Affinity purification of molecules is described in Chapter

5 and the preparation of immunoglobulins (for consideration as immunogens in this case) is described in Section 10.

The following are some examples of immunogen preparation, and approaches to immunization, which may be of special value.

### 3.2.1 *Use of immune complexes*

An immune precipitate often provides a powerful stimulus for antibody production representing a natural form of immunization. Precipitates containing only a few ng of immunogen can induce a good antiserum when used repeatedly with adjuvant.

(i) *Test tube precipitates.* Given a small volume of a specific starting antiserum the immunogen can be selectively precipitated from a solution of mixed molecules in the test tube. Precipitation is assisted by the presence of 3% (w/v) polyethylene glycol (PEG) (6000 mol. wt—see Appendix). The purified IgG fraction of the antiserum is recommended to avoid complement components in the complex. The best approach is to perform a number of reactions in small precipitin tubes with constant antibody and a range of immunogen mixture concentrations added in equal volume in PBS (0.01 M, pH 7.4) with 6% PEG, which gives visible precipitation in the middle range after incubation for 1 h at 37°C and overnight at 4°C. Select for immunization a precipitate in mild antigen excess, centrifuge at 10 000 *g* for 1 h and wash twice in 3% PEG–PBS. The precipitate is finally suspended in a small volume of PBS and prepared in FCA for injection (see Section 4). Naturally the animal will also respond with antiglobulin antibodies unless the same species is used for both the original and second antibody response. Antiglobulins can, however, be easily removed by absorption (see Section 9).

PEG precipitation of immune complexes from serum may be a useful way of securing for immunization antigens suspected of circulating as soluble complexes in the blood. Equal volumes of serum and 4% PEG are mixed and incubated for 1 h at 4°C. This is then centrifuged at 10 000 *g* for 1 h at 4°C and the precipitate washed twice in 2% PEG–PBS. This is then resuspended in PBS and used with FCA.

(ii) *Precipitation lines and peaks in gels.* The use of excised carefully washed and homogenized gel precipitation reactions is a simple method of obtaining a unispecific reagent where only a multispecific antibody was previously available and, as with test tube precipitates, making an antiserum that copies an original in short supply without the need for expert and laborious immunogen separation. In 2D-IEP (see Chapter 6) complex antigen mixtures previously separated in one dimension are migrated electrophoretically into a layer of agarose gel with its respective antibodies. In parts of the gel where any such peak does not overlap with another it can be carefully excised and used as purified complex for immunization. The method is suitable for any antigen (immunogen) molecule that migrates anodally at pH 8.6 and is present in the starting mixture at >10–20 μg/ml, to form a visible peak. An example of this approach is shown in *Figure 2*. To work satisfactorily the following advice may be useful.

(1) Ensure that electrophoresis in the first dimension is optimal to achieve the best separation of the target antigen peak from other peaks.
(2) Use an electrophoresis tank with large buffer reservoirs; use fresh electrophoresis buffer and new or thoroughly washed lint wicks so that the agarose is not contaminated with previously separated proteins.

(3) Use the IgG fraction of the antiserum in the agarose to reduce agarose protein contamination, and adjust the concentration so that the chosen precipitation peak is fully formed and stands clear of other peaks.
(4) Run the gel, on a cooling plate, overnight or longer to ensure that non-precipitated molecules are migrated out of the gel.
(5) Run several plates so that several peaks can be prepared, for initial and boosting injections, at the same time.
(6) Cut out the peaks and soak these in saline (20 ml in a glass universal bottle) at 4°C with several changes for at least 1 week. One or more can then be homogenized in a small volume of PBS using a glass tissue homogenizer or electric blender and then emulsified with FCA.

In the use of antigens that migrate cathodally at pH 8.6, precipitation lines can be excised from IEP reactions developed with multispecific antisera. Only those parts of lines that do not cross or merge with other lines should be used. Use of the IgG fraction of the antiserum is strongly advised as well as very extensive washing.

Using a unispecific antiserum, individual antigens from a mixture can be precipitated in gel using RIEP (Chapter 6). The advantage of the rocket approach is that numerous identical rockets can be prepared on the one plate. The same measures should be applied as for two-dimensional peaks, but in this case the first dimension pre-run is not required. However, the method is sensitive in revealing any contaminating antibody in the so-called unispecific reagent that reacts to other components of the antigen mixture, so ensure that only the desired part of the rocket is excised for immunization.

3.2.2 *Use of nitrocellulose bands and blots*

Sodium dodecylsulphate–polyacrylamide gel electrophoresis (SDS–PAGE) has become a routine practice for demonstrating complex molecular mixtures. The capacity to transfer electrophoretically the separated component bands onto nitrocellulose paper (Western blotting) not only allows very sensitive tests for the discrete specificity of antibody (which defines the molecular size of the target antigen and allows comparison of antibodies) but also has a newly recognized potential for the harvesting of separated macromolecular bands for immunization. It has been shown that bands revealed by protein staining, excised and ground to a fine powder, retain antigenic integrity. There is one report that mice can be immunized with as little as 10 ng of protein adsorbed to nitrocellulose if this is implanted in the mouse (22). Larger amounts of antigen adsorbed to nitrocellulose have been used for immunization after reduction to small particles in dimethylsulphoxide (23).

3.2.3 *A simple approach to raising antiglobulin sera using red cells*

The response of an animal to the red cells of another species is to give both IgM and IgG antibodies. Anti-red cell antibodies will then coat a suspension of target red cells which when injected into the red cell donor will induce an antiglobulin response. This simple approach avoids the necessity for immunoglobulin purification and produces excellent antiglobulin reagents. As an example, antibodies to sheep red cells (SRBC) can easily be raised in rabbits by injecting intravenously 1 ml of 10% (v/v) saline-washed sheep erythrocytes. Bleeding 10 days after the first injection will yield predominantly

IgM antibodies. Re-injecting a week later and twice weekly for several weeks, bleeding 2–5 days after the last injection, will yield mostly IgG antibody. These sera can then be mixed.

(i) Heat 0.5 ml of mixed serum to 56°C for 20 min to destroy complement binding. Add to 5 ml of washed 10% red cells in a glass bijou bottle, cap it and mix rapidly by inversion.

(ii) Slowly turn the cells in a rotary mixer for 20 min and then wash them five times in cold PBS.

(iii) The washed, antibody-coated cells can then be used to immunize sheep.

A similar approach can be used to produce, for instance, anti-mouse immunoglobulin sera in the rabbit. Immunization procedures are covered in Section 5.

Separate anti-IgM and anti-IgG sera can be raised by a modification of this approach. In this case the rabbit's early response serum is taken, the IgM separated from IgG and used exclusively to coat cells for immunization. Later responses are used to separate the IgG for coating. The preliminary stages are as follows:

(i) *IgM separation.* 5 ml of heat inactivated serum is applied to a 40 × 3.5 cm Sephadex G200 column in PBS, pH 7.2 and passed through at 25 ml/h, collecting 5 ml fractions. Each fraction is tested for agglutinating activity and the first four positive samples [leading edge of the first (IgM) eluted peak] are pooled and used to coat red cells for immunization.

(ii) *IgG separation.* 5 ml of inactivated serum from repeatedly injected rabbits is also separated by Sephadex G200. In this case the four eluate samples representing the centre of the second (IgG) eluted peak, with agglutinating activity, are collected, pooled and used for red cell coating.

In our experience antisera raised in this way have only minor (anti-light chain) cross-reactivity which can be absorbed out with the antibody-coated cells of the other class.

### 3.2.4 *Tolerance induction*

Antibodies that recognize minor structural differences between related molecules of the same species are often difficult to produce in isolation in another species where the tendency is to respond predominantly to the broader antigenic differences existing between donor and recipient. It is sometimes possible to induce specific tolerance towards the latter category of epitopes and thus to change the emphasis of the antibody response. This can be applied both to polyclonal and Mab production. We have successfully raised antisera in sheep to all the *a* and *b* locus allotypes of rabbit immunoglobulins using the ploy of injecting, intravenously, newborn lambs with 500 mg of pooled ultracentrifuged, purified rabbit IgG that lacks a single allele, and at 3 months injecting this allele in FCA intramuscularly.

In the production of Mabs to the human IgG subclasses, where again only minor structural differences exist in regions determining subclass antigenicity, it has been found useful to tolerize mice by intraperitoneal injection of approximately 12 mg of some human IgG subclass paraproteins a week before immunization with other subclasses (24). The tolerizing paraproteins were biofiltered by passage through donor mice whose serum was then used to supply the tolerogen.

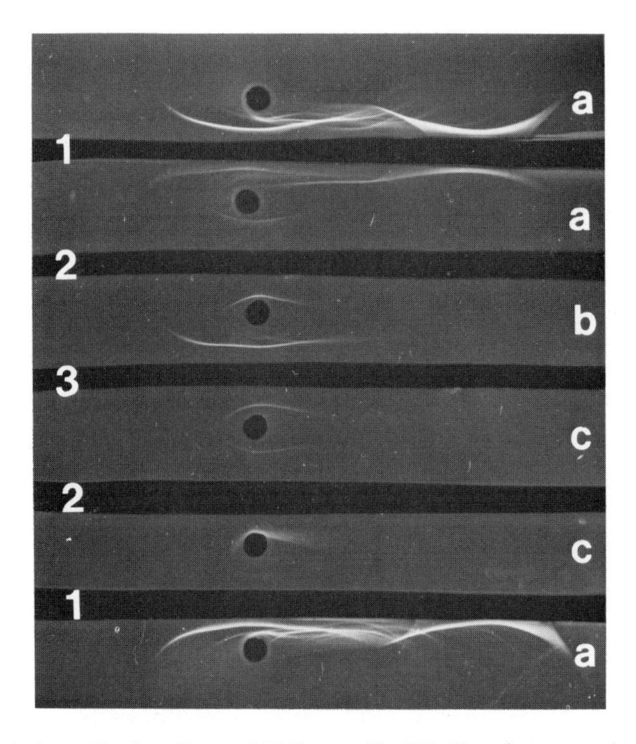


**Figure 1.** Steps in the purification of human IgM illustrated by IEP. The reference reaction of whole serum with anti-whole serum shows the range of possible contaminants, the reaction with oligospecific anti-Ig shows contamination of the IgM with other Ig classes in the impure preparation; both antisera confirm the purity of the final product, whose position is demonstrated by the IgM-specific reference reagent. Wells: (**a**) contains whole normal human serum, (**b**) contains partially-purified IgM, (**c**) contains purified IgM. Troughs: (**1**) contains polyspecific anti-whole human serum, (**2**) contains reference anti-human IgM serum, (**3**) contains reference anti-human immunoglobulins (IgG, IgA, IgM) serum.

### 3.2.5 *Modulation of the antibody response to imunogens by complexing with antibody to some epitopes*

In some cases it may be possible to change the emphasis of antibody responses to multi-determinant immunogens by attaching antibodies (i.e. Mabs of defined epitope specificity) and using the complex for immunization. It has recently been shown that the response of mice to a human IgGλ (i.e. an IgG with λ light chain) contains high titres of anti-λ antibody when the molecule is complexed with one or more anti-gamma chain Mabs, but the anti-λ response is depressed using an Mab to the λ chain (25). Furthermore, complexing the IgG with an anti-gamma chain (Fcγ)-specific Mab enhanced the response to the poorly immunogenic (γ1) domain, whereas no anti-Cγ1 response was produced in mice receiving the IgG complexed with an anti-Cγ1 Mab. Similar suppressive effects could not be obtained after coupling antibody to strongly immunogenic regions of the molecule. This approach clearly has importance in strategies for directing the specificity of antibody responses towards parts of molecules with poor or moderate immunogenicity.

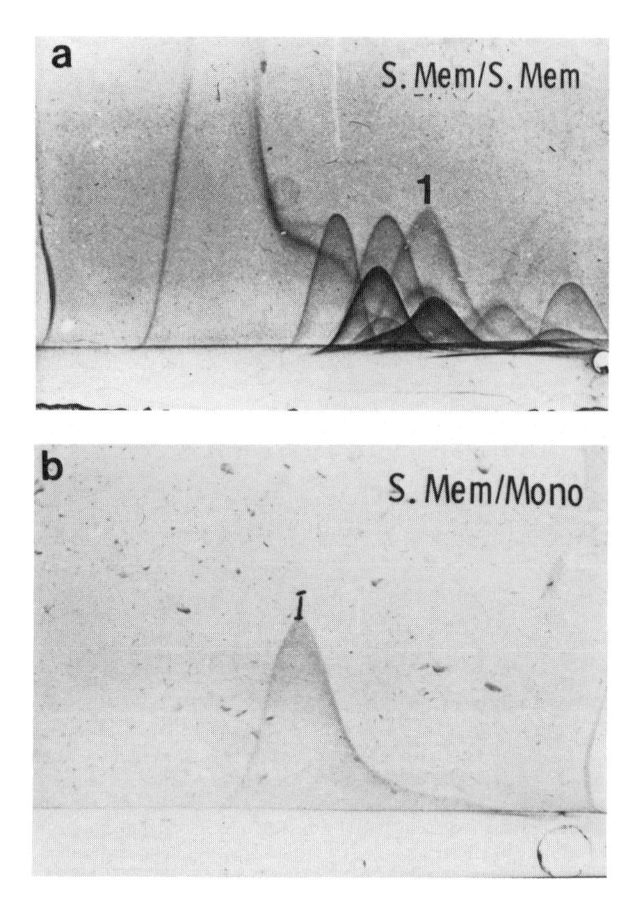


**Figure 2.** 2D-IEP of *S. mansoni* adult worm surface membrane soluble antigen extract. (**a**) The reaction using a polyspecific antiserum in the agarose raised by chronic, long-term immunization of rabbits with low doses of the antigen mixture in FCA; (**b**) The reaction using the same antigen mixture but with a unispecific antiserum in the agarose raised by injecting the excised precipitation peak of antigen **1** (position shown in **a**).

## 3.3 Testing the purity and complexity of soluble immunogens

In preparing unispecific antibodies most of the problems arise not in obtaining a good antibody response but in absorbing out unwanted specificities arising through the presence of contaminants in the immunogen. Thus it is important to check purity and identify contaminants. Even if these cannot be removed a knowledge of their nature prepares the ground for subsequent antiserum absorption steps. Where the intention is to raise polyspecific antibodies it is equally important to demonstrate the molecular complexity of the component mixture as this gives a reference point in determining success in achieving a balanced diverse antiserum to the constituents.

### 3.3.1 *Antibody methods*

This is based on the use of a polyspecific antiserum which demonstrates either the purity of an isolated immunogen compared with starting material or the quality of a reference

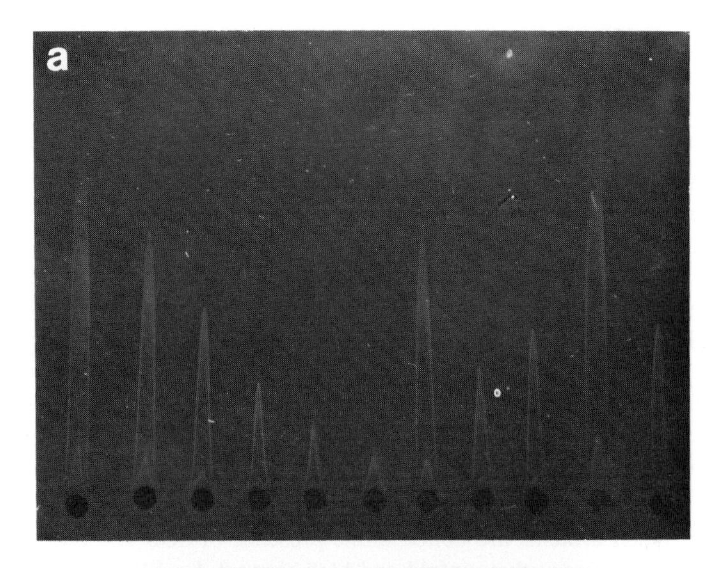


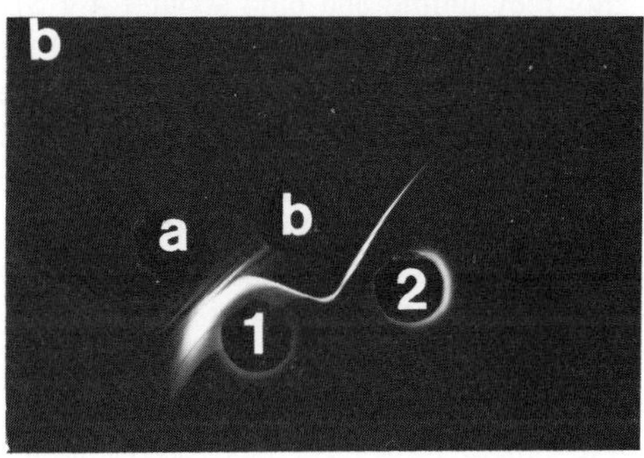


**Figure 3.** Examples of two gel precipitation methods to test for purity of immunogen by using antibodies. (**a**) RIEP test for purity of serum albumin (in wells) run into agarose gel containing anti-whole serum protein antibodies. The taller rockets are the albumin reaction but inside these can be seen a reaction to a contaminating antigen at lower concentration which gives smaller rockets. (**b**) GDD reaction to test the purity of albumin: well **1** contains a polyspecific antiserum to serum proteins, well **2** contains a unispecific anti-albumin serum. Well **a** contains an impure albumin preparation which gives multiple precipitation lines against the polyspecific reagent; well **b** contains a pure albumin preparation which gives a single line against the polyspecific reagent, and this is confirmed as albumin by reaction of identity with the reference anti-albumin reagent.

reagent to be aimed for in the preparation of a new antiserum of equivalent broad reactivity.

Useful tests in this category are gel double diffusion (GDD) (the least sensitive), IEP (for single or complex antigens), RIEP (for single antigens) and 2D-IEP (for single and complex antigens). The methods for these tests are described in Chapter 6. *Figure 1* illustrates the purification of human IgM from whole serum as shown by IEP using a reference antiserum against the major components of the starting mixture (anti-human whole serum) and a second specific reference serum to the molecule under purification (anti-human IgM—$\mu$ chain-specific) to identify positively the prepared molecule. *Figure 2(a)* illustrates the 2D-IEP distribution of *Schistosoma mansoni* adult antigens achieved with a reference polyspecific antiserum. Using such an antiserum the purity of single antigen components may be adequately demonstrated. *Figure 3* demonstrates how purity may be tested by GDD and RIEP.

### 3.3.2 *Electrophoretic separation of proteins in gels—the SDS–PAGE technique*

A major method for the analytical separation of proteins by gel electrophoresis, which is now widely used as a standard technique, is SDS–PAGE. This demands specialized equipment and costly reagents but is indispensable to those engaged routinely in production and application of antibodies. The method is valuable not only because of its molecular resolving power over a wide molecular size range but also because it allows, through additional steps, the definition of antibody specificity. It is not possible to provide here a full account of all variations in this technique—the reader is referred to an excellent book in this series (26). What follows is a minimum description of the principle and simplest methodology applied to the analysis of proteins by this method. The additional steps required to use the system for tests of antibody specificity are considered in Chapter 6.

(i) *Principle.* When proteins are boiled in the presence of SDS and reducing agents such as 2-mercaptoethanol they unfold and bind about 1.4 g SDS/g of protein. The dissociated constituent polypeptide chains assume a conceived rod-like state in which the diameter of the rod is thought to be constant whilst the long axis varies in proportion to the molecular weight. The binding of SDS results in a uniform negative charge to polypeptide chains and in consequence they assume a constant charge:mass ratio. When, by virtue of this charge, they are electrophoretically migrated through the supporting matrix of polyacrylamide, their mobility is inversely proportional to the logarithm of the molecular weight. Thus at the end of the run small molecules have run farthest along the tracks in the gel, larger molecules the least and each constituent, after staining for protein, can be compared with molecular size markers run in a reference track, and molecular sizes (kd) assigned. To achieve maximal resolution of bands the protein mixtures are applied to a narrow upper band of stacking gel (using a comb template to prepare the tracks). The stacking gel, with less acrylamide and minimal sieving properties, creates with the sample buffer a discontinuous buffer system of chloride and glycinate ions at pH 6.8 in which a moving boundary with leading chloride ions gathers the proteins into a narrow contained zone before they enter the resolving gel.

Polyacrylamide is a polymerization product of acrylamide and cross-linkers such as *N,N′*-methylene bisacrylamide (BIS). The polymerization is initiated by a catalyst–redox system such as ammonium persulphate–*N,N,N′N′*-tetramethylene diamine (TEMED) or riboflavin–TEMED, the former being the more popular for gels of 3–4% acrylamide. The high chemical and mechanical stability of polyacrylamide gels makes them ideal media for protein electrophoresis.

(ii) *Reagents.* See *Table 3* for a list of reagents.

The following points should be noted:

(1) Acrylamide and BIS are neurotoxins especially in the unpolymerized state, but polymerized gels may also contain some unreacted monomer. Wear a mask and gloves to weigh out the reagents, preferably in a fume cupboard and always wear gloves to handle the solutions.

**Table 3.** Reagents used in the electrophoretic separation of proteins in gels.

| | |
|---|---|
| *Acrylamide stock solutions* | |
| *Stock 1* Acrylamide 44 g | made up to 100 ml with distilled water |
| BIS 0.8 g | |
| *Stock 2* Acrylamide 30 g | made up to 100 ml with distilled water |
| BIS 0.8 g | |
| [Filter through Whatman No. 1 filter paper]. | |
| *Sample disruption buffer* | |
| 10% (w/v) SDS | 5 ml |
| 0.5 M Tris–HCl pH 8.8 | 2.5 ml |
| Distilled water | 5 ml |
| Glycerol | 2.5 ml |
| 2-Mercaptoethanol | 0.25 ml |
| 5% (w/v) Bromophenol Blue | 0.2 ml |
| *Separating gel buffer pH 8.9* | |
| Tris | 36.6 g |
| 1 M HCl | 48 ml |
| Distilled water to 100 ml | |
| *Stacking gel buffer* | |
| Tris | 12.1 g |
| Conc. HCl to pH 7.0 | |
| Distilled water to 100 ml | |
| *Other gel components:* | |
| SDS solution 10% (w/v) in distilled water. | |
| Catalyst: ammonium persulphate 10% (w/v) in distilled water—prepare fresh just before use. | |
| TEMED—use as purchased. | |
| *Electrode (running) buffer (pH 8.3)* | |
| Tris | 30 g |
| Glycine | 144 g |
| SDS | 10 g |
| Distilled water to 10 litres | |
| *Staining solution* | |
| Methanol:acetic acid:water (25:10:65 by vol). | |
| 0.03% (w/v) Coomassie Brilliant Blue | |
| Filter before use. | |
| *Destainer* | |
| Above solvent without stain. | |

(2) Stock solutions can be stored for about 4 weeks at 4°C but should be discarded if any cloudiness is present. Protect acrylamide solutions from light. SDS solutions should be stored at room temperature. This includes the electrode buffer.

(3) Ammonium persulphate and TEMED are stored at 4°C.

(4) Gel solutions should be degassed for 10–15 min before introducing the catalyst.

**Table 4.** Preparation of 14 cm width separating gel of 3 mm thickness.

| | Separating gel (ml) | | Stacking gel (ml) |
|---|---|---|---|
| 14% gel | | | |
| Stock 1 | 18.75 | Stock 2 | 5.0 |
| 10% SDS | 1.50 | | 0.3 |
| 1.5 M Tris–HCl pH 8.8 | 18.75 | | 7.5 |
| Distilled water | 20.0 | | 16.0 |
| TEMED | 0.140 | | 0.08 |
| 10% ammonium persulphate | 0.20 | | 0.10 |
| Time to set | 15 min | | 20 min |
| 10% gel | | | |
| Stock 1 | 22.2 | Stock 2 | 2.0 |
| 10% SDS | 0.66 | | 0.1 |
| 1.5 M Tris–HCl pH 8.8 | 16.6 | | 2.5 |
| Distilled water | 25.18 | | 5.3 |
| TEMED | 0.066 | | 0.02 |
| 10% ammonium persulphate | 0.66 | | 0.10 |
| Time to set | 15 min | | 15 min |

(iii) *Preparation of gels.* For 14 cm width separating gel of 3 mm thickness, see *Table 4*.

(iv) *Casting gels.* Many gel kits are available—they all provide pairs of glass plates, Teflon combs for forming wells in the stacking gel, spacers for determining the gel thickness, cassette and gel casting platforms, clamps and crams. An electrophoresis (vertical) tank, power pack, gel washing and staining dishes and drying kit are the other requirements. Sources of equipment and reagents are given in the Appendix.

(1) Wash the glass plates, combs and spacers, and clean them with methanol to remove any grease. Form the cassette by fitting the spacers (we use 3.0 mm) between the plates at the base and two sides and clamping the unit together. Place the unit upright on the pre-greased rubber-padded gel casting platform and cram it into the support to provide a good seal. With some apparatus it may be necessary to pre-seal the bottom of the cassette with 1% (w/v) agarose.

(2) Prepare the separating (lower) gel mixture of desired concentration and mix; add and mix the catalyst just before pouring.

(3) Pour the gel solution smoothly but rapidly into the cassette by means of a syringe or pipette, exercising care that air bubbles do not become trapped. Immediately after pouring, overlay the gel with water to exclude air and enhance polymerization. This helps also to level and smooth the top of the gel. Gels may also be overlaid with 0.1% SDS in water, isoamyl alcohol or *n*-butanol—these latter two are less dense than water and there is a reduced risk of mixing. However it is better to overlay with water or separating buffer if the gel is to be kept overnight. Allow the gel to polymerize for at least 30–45 min (overnight recommended).

(4) Prepare the stacking gel mixture. Before adding the catalyst, remove the lower gel overlay and rinse with water; add catalyst, mix and pour the stacking gel.

Immediately insert the comb of selected size to form the wells. Overlay the stacking gel with water to prevent evaporation and leave to set for 20−30 min.

(5) Meanwhile prepare samples in disruption buffer (equal volumes of sample and buffer) and place them in a boiling water bath for 2−5 min. Samples should contain approximately 2−5 mg/ml concentration of protein, using about 50 μl (100−250 μg) per well. Ideal concentrations vary according to the complexity of the sample and can only be determined by trial.

(6) Remove the comb from the stacking gel and wash the wells thoroughly with 0.1% SDS in 0.5 M Tris−HCl three times to remove acrylic acid formed during polymerization.

(v) *Running the gel*

(1) Load the wells with the disrupted samples—the height of the loaded sample in each well should be less than the distance from the bottom of the well to the top of the separating gel. Fill the wells with running buffer.

(2) Fill the electrophoresis buffer tank with running buffer and turn on the cooling water circulation (when present). Place the gel cassettes in the buffer tank and connect the leads.

(3) Apply current at 10−15 mA per gel for about 1.5 h and then double to 20−30 mA for a further 1−2 h until the dye front is about 1 cm from the end of the plate. Turn off the current, disconnect the leads and remove the cassette.

(vi) *Staining and drying gels*

(1) Open the cassette and gently remove the gel under water in a shallow dish. Immerse the gel completely in protein staining solution for 1 h or more.

(2) Transfer the gel to destainer solution and leave it overnight until the background staining is removed. Enhanced staining can be achieved by the silver staining method for which kits are available (see Appendix). This is particularly useful in circumstances of low starting protein concentration and to bring out minor protein component bands.

(3) For drying, place the gel on a smoothed layer of polythene clingfilm and cover with filter paper trimmed to the appropriate size. Place the gel on the vacuum gel dryer. Dried gels should be stored in pressed condition—i.e. between the leaves of a book, to prevent wrinkling.

The above method was proved to be applicable in the separation of bacterial- and parasite-extracted antigens and the components of human sera, giving good banding of molecules in the 150−10 kd range. Changes in acrylamide concentration can be made to provide resolution in more restricted molecular size ranges. In addition minigel kits are available for shorter runs (see Appendix) which include their own gel forming cassettes. These are more economical of reagents and can be run in a shorter time but at the expense of resolution of components in very complex mixtures. The compromise is to run minigels with several acrylamide concentrations. *Figure 4* illustrates the resolving power of a 14 cm width 14% gel for the membrane-extracted, disrupted proteins of *Escherichia coli* and *Mycobacterium* species. The latter is the starting material used for the production of both Mabs and polyspecific antisera in rabbits. The preparation

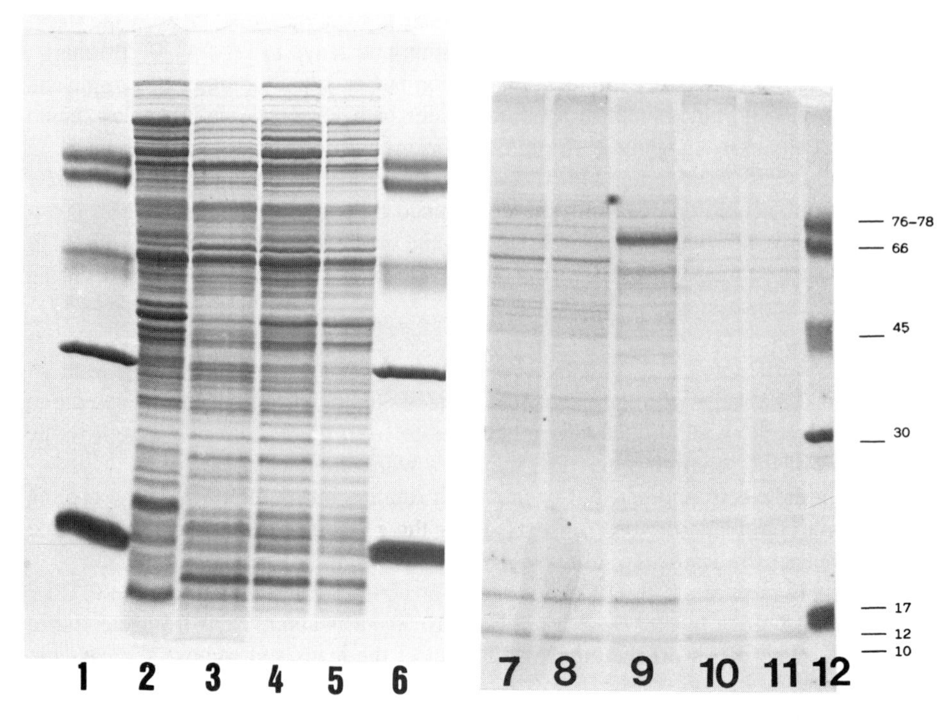


**Figure 4.** SDS–PAGE separations of *E.coli*, *M.bovis* BCG and *M.tuberculosis* membrane extracts in a reducing 14% gel. The plates are stained for protein bands. **Tracks 1, 6** and **12** are mol. wt. markers as indicated. **Tracks 2–5** are *E.coli* inner and outer membrane proteins, **7** and **8** *M.tuberculosis* and **9–11** *M.bovis* BCG membrane antigens and total soluble extract.

of the *E.coli* proteins was as follows (27):

(1) A 250 ml culture of *E.coli* in mid-log phase was centrifuged at 4°C for 10 min at 10 000 *g* and the pellet resuspended in 25 ml ice-cold 50 mM PBS, pH 7.4. This was again centrifuged and the pellet now resuspended in 2 ml of PBS with 140 mM 2-mercaptoethanol (2-ME) in a small glass bottle.

(2) The suspension was sonicated on ice at maximum power using a Life Science Laboratory W380 Ultrasonic 1 cm probe at maximum power (380 Watts) using an 8 × 30 sec pulse/rest cycle to allow cooling in the ice bath.

(3) The sonicate was centrifuged at 10 000 *g* for 10 min at 4°C and the supernatant ultracentrifuged at 100 000 *g* for 1 h at 4°C in a pre-cooled rotor. The supernatant constitutes the cytosol.

(4) The pellet was resuspended on 1 ml PBS (without 2-ME) and recentrifuged for 30 min at 100 000 *g* at 4°C and the pellet resuspended in 100 μl of PBS.

(5) After 30 min at room temperature 10 μl of 20% (w/v) sarcosyl (sodium *N*-lauryl sarcosinate) was added to solubilize the membrane proteins. The mixture was again centrifuged at 100 000 *g* for 30 min at 10°C. The supernatant constitutes the inner membrane proteins, the pellet the outer membrane proteins.

*Figure 5* illustrates a test for imunogen purity using SDS–PAGE.

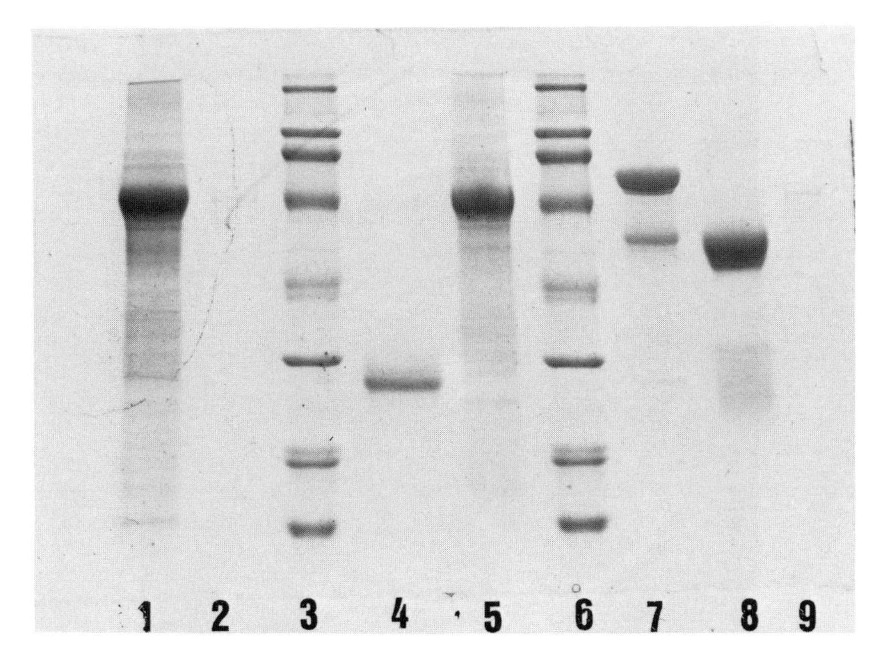


**Figure 5.** SDS–PAGE used to determine purity of immunogens. **Left—Tracks 1–3**, mol. wt markers, BCG membrane extract and the 65 kd antigen. **Right—Tracks 1–4**, mol. wt markers, whole human serum, Ig light chain and IgG heavy chain. The purified BCG antigen was kindly provided by Dr Coulson from the National Institute of Medical Research, Mill Hill, London, and the immunoglobulin subunits by Dr M.Walker.

## 4. PREPARATION OF ADJUVANTS FOR IMMUNIZATION

### 4.1 Water-in-oil emulsions

#### 4.1.1 *Freund's Incomplete Adjuvant*

The oil is prepared by adding 35% (v/v) Arlacel A emulsifier (mannide monoleate) to Bayol F light paraffin oil. This is thoroughly mixed and can be stored indefinitely at 4°C.

(i) *Equipment required to prepare the immunogen emulsion*

2 ml graduated pipettes with strong bulb
5 ml glass bijou bottle
1 or 2 glass Luer-lock syringes with Vaseline-lubricated pistons
20-gauge 1 in needle [fused 20-gauge steel (double hub) needles or Teflon double-ended syringe adaptor with threaded points (see Appendix) are desirable alternatives]
50 ml beaker of tap water
1 ml or 2 ml glass syringes for injection with Vaseline-lubricated pistons
Injection needles

(ii) *Methods*

(1) Select the appropriate total volume of emulsion required according to species, site(s) of injection and number of animals. Add 20% for wastage.

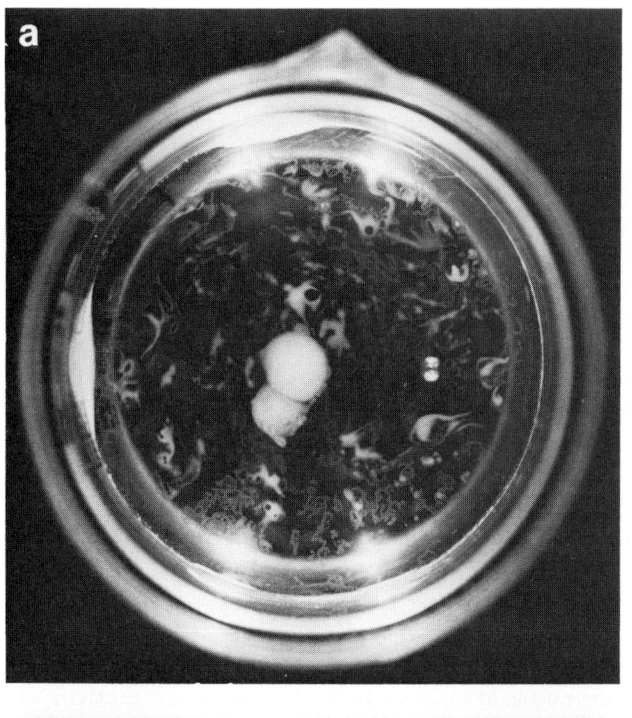


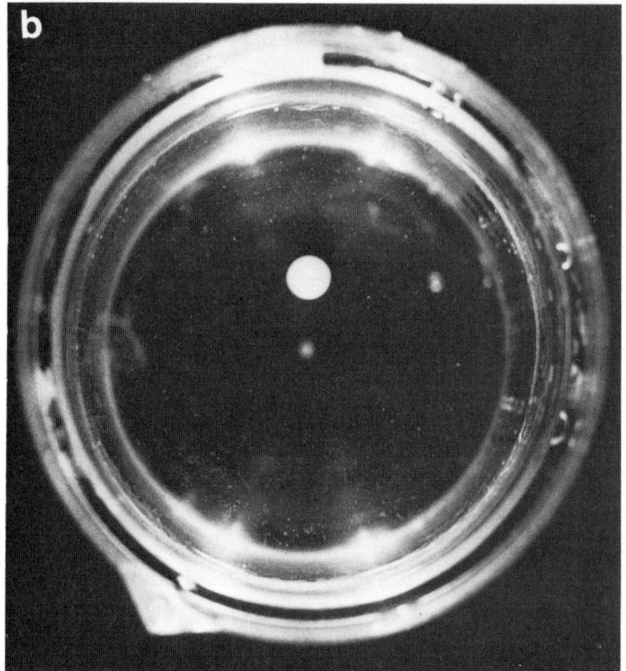


**Figure 6.** Water flotation test for stability of water-in-oil emulsion of immunogen. (**a**) is the result with an unstable, inadequately mixed emulsion which disperses; (**b**) is a stable emulsion which floats as a discrete droplet.

(2) Pipette the appropriate volume of adjuvant into a bijou bottle.
(3) Prepare an equal volume of the immunogen solution (or fine suspension or precipitate as relevant) in aqueous phase and draw into one syringe, excluding all air, using the 20-gauge needle.
(4) Inject the aqueous immunogen phase vigorously into the oil phase and then draw the complete volume carefully into the syringe without addition of air. Using the double-hub or double-ended adaptor system, remove the single needle from the filled syringe and attach the adaptor, push air out gently and attach a second lubricated glass syringe to the other port. Ensure both attachments are screwed tightly and then work the emulsion from one syringe to the other many times until the white emulsion becomes stiff and difficult to work further. Using the retained single needle system, pass the emulsion from the syringe into the bijou and back repeatedly without taking up air, again until the emulsion becomes hard to work.
(5) A stable water-in-oil emulsion is not formed until a drop allowed to fall from the needle onto clean cold water sinks briefly as a discrete drop and then floats without any dispersion (*Figure 6*). If the drop disperses the emulsion should be worked further and tested again in the refilled beaker. This is a very important step as an unstable emulsion when injected will be rapidly dispersed in the tissues.
(6) Return the completed emulsion to the bijou bottle and withdraw into 1 ml or 2 ml glass, lubricated, syringes, each with the required volume of emulsion for a single site (or single injection). Attach and mask the appropriate-sized Luer-lock needle. It is a good idea to place this in a labelled wrapper—or at least ensure that the syringe is labelled and the needle firmly masked in transit. Ensure that the needle is very firmly attached as considerable pressure may be needed to inject. Methods for injection with adjuvant are detailed below (Section 5.1.4).

#### 4.1.2 *Freund's Complete Adjuvant*

(1) Starting with a dry heat-killed powder of *M.tuberculosis* (H37Rv/or H37Ra) or *M.butyricum*, prepare a stock suspension at 10 mg/ml by grinding the organisms in a small volume of Bayol F in a pestle and mortar.
(2) Add the suspension to FIA at a concentration of 0.5 mg original dry wt per ml of the prepared FIA mixture—that is, 1 ml in 20 ml to prepare FCA.

The suspension should be thoroughly mixed before use, especially after storage (4°C) when the organisms settle. FCA can be sterilized by autoclaving or irradiation and will store indefinitely. Preparation of FCA emulsions follows exactly the protocol described for FIA above.

**Warning**: Accidental injection or pricking of the skin with FCA may cause a prolonged and very painful local anti-mycobacterial inflammatory reaction. Every precaution should be taken to avoid such accidents. Bleeding should be encouraged at the site and medical attention always sought even though the immediate damage may appear minimal.

### 4.2 **Alum-precipitated proteins**

The following method is recommended for preparing an alum precipitate of serum albumin or IgG at a concentration of 10–20 mg/ml (28).

(i) Add 4.5 ml of 1 M $NaHCO_3$ at 20°C to 10 ml of protein and mix.
(ii) Add 10 ml of 0.2 M aluminium potassium sulphate ($KAl[SO_4]_2.12H_2O$) in distilled water slowly with stirring. Hold the mixture at 20°C (room temperature) for 15–20 min and then centrifuge at 3000 *g* for 10 min.
(iii) Wash the precipitate three times with PBS before use. It can be stored with 0.01% w/v thiomersal preservative.

If required, for other proteins, a more detailed procedure has been described (29). Alum precipitates are commonly used as intramuscular or intraperitoneal injections but can be used in conjunction with FCA or FIA to form the water phase of the emulsion.

### 4.3 Bacterial suspensions as adjuvants

A *B.pertussis* killed suspension at $10^{10}$ organisms/ml is used commonly as an adjuvant in rodents. Usually this is in association with alum-precipitated protein, the two mixed and injected intraperitoneally, using $2 \times 10^9$ bacteria/mouse injection. Both *B.pertussis* (in mice) and *Proteus vulgaris* $\phi$X19 (in rabbits), as heat-killed suspensions, have been used to raise anti-bacterial antibody for the subsequent antibody-coating of organisms used in the autologous species to raise anti-Ig allotype and anti-idiotype sera. Further details of these applications are given as examples of immunizations in Sections 6.5 and 6.6.

### 4.4. Liposomes

The following method for preparing immunogen-containing liposomes is from refs 8 and 9.

(i) Dissolve 20 mg of sphingomyelin and 8 mg of cholesterol in 3 ml of chloroform and add 3 ml of ether.
(ii) Bath sonicate the mixture to a uniform water-in-oil emulsion with 10 mg of immunogen dissolved in a balanced salt solution (BSS) of 5.5 mM glucose, 0.4 mM $KH_2PO_4$, 1.2 mM $Na_2HPO_4.7H_2O$, 1.3 mM $CaCl_2.2H_2O$, 5.4 mM KCl, 136 mM NaCl, 1 mM $MgCl_2.6H_2O$, 0.8 mM $MgSO_4.7H_2O$.
(iii) Make the liposomes up to 2 ml with BSS and treat with 0.1% osmium tetroxide (final concentration) in saline.
(iv) Hold the suspension for 30 min at room temperature and then dialyse it extensively against a continuous flow of water.
(v) Separate the liposomes from free immunogen by centrifugation at 1000 *g*. Measure the protein incorporation by the Lowry method.
(vi) For injection, make up the liposome suspension in BSS to appropriate incorporated-immunogen concentration.

Carpet viper (*Echis carinatus*) venom in osmicated liposomes has been injected as 0.2 ml aliquots intravenously into mice to obtain a persistent high titre anti-venom antibody response for the lifespan of the mouse (30). Similar prolonged responses are observed after single intravenous or subcutaneous injections in rabbits and intravenous injections in sheep.

## 5. IMMUNIZING AND BLEEDING ANIMALS

### 5.1 **Injections**

#### 5.1.1 *General points*

(i) Whatever the route, always choose the minimum required syringe size for a single site or dose injection; use a new disposable, or sterilized glass, syringe and a new needle, of appropriate gauge and length, which is firmly attached.

(ii) Always sterilize the injection site with an alcohol swab and never inject into a site obscured by hair/fur/wool—this should be cut back with clippers, scissors or shaved with a scalpel blade to give sight of the skin and superficial veins. Swab the injection site after needle removal and check that no sustained bleeding occurs.

(iii) Ensure that the animal is firmly held with minimum stress—use an assistant where necessary and never attempt injections alone without previous training.

#### 5.1.2 *Specific points*

(i) Always use a glass syringe and Luer-lock needle attachment for injecting adjuvant emulsions.

(ii) For intramuscular injections, after penetration to the required depth, withdraw the piston a little to ensure that the needle has not entered a blood vessel.

(iii) Avoid superficial blood vessels.

(iv) Be very cautious about repeated intravenous injections of soluble immunogen as many species are susceptible to acute systemic hypersensitivity reactions.

(v) Footpad injections with depot adjuvants cause unnecessary pain and disability and should not be practised—equivalent responses can be gained from other depot locations.

(vi) Using FCA be very careful that no emulsion follows the track of the needle on withdrawl and is left in the superficial tissues. This will lead to the development of open skin lesions which cause pain and are easily infected.

#### 5.1.3 *Routes of injection*

Animals may be immunized by intradermal (i.d.), subcutaneous (s.c.), intramuscular (i.m.), intraperitoneal (i.p.) or intravenous (i.v.) routes. The choice of route may be influenced by the physical nature of the injection, the species and the stage of the immunizing routine.

(i) Intradermal injections are seldom used to initiate a response but may be used for boosting with immunogen in saline—especially in rabbits and guinea pigs. The associated developing cutaneous Arthus reactions allow a slow release of *in situ*-formed immune complexes into local draining lymph nodes.

(ii) Subcutaneous and i.m. routes are used for immunogens in Freund's adjuvants as they favour granuloma formation round depot sites and slow immunogen release to local lymph nodes.

(iii) Alum precipitates can be injected by all routes although usually given i.m. or i.p. and rarely i.v.

(iv) Cell and bacterial suspensions, liposome suspensions and immunogens in solution are usually administered i.p. or i.v. Substantial amounts of i.p. injected material (even cells) reach the circulation rapidly and hence involve the spleen in the response. The i.p. route may prevent acute hypersensitivity death in repeated injections of immunogen in solution—even by this route, however, injections should be given slowly or in repeated small aliquots over several hours or days. FCA should not be administered i.p. as this causes a generalized peritoneal inflammation.

### 5.1.4 *Injecting different species*

#### (i) *The mouse*

*Intravenous injections.* This is the most difficult of common laboratory species injections, especially in dark-skinned strains and requires considerable practice. Training should be done with 0.1% (w/v) Evans Blue in saline added to the injection. Injection is made into the mid dorsal tail vein. The mouse is held in a restrainer such as a modified 60 ml plastic syringe with the protruding tail first immersed in warm (not scalding) water to dilate the vein for about 2 min. The restrainer is then clamped at a convenient tail-downward angle and the tip of the tail held with the left thumb and second finger, the index finger being used to support the tail from beneath. The immunogen is already prepared in a 1 ml syringe. Ensure that no air is present and that the piston moves freely. Use a 10 mm, 25-gauge needle. Swab the supported tail surface with alcohol. With the needle point on the lower side gently introduce the needle into the vein with the right hand, at an angle closely parallel to the tail. The vein is very superficial and correct insertion offers little resistance. Run the needle about 5 mm into the vein and inject using the thumb or palm of the hand. Inject a little volume slowly to begin with to check that the fluid runs up the vein. If resistance is felt and a bleb appears then the needle is not properly located. Injection can still be achieved by moving up the tail. Volumes up to 0.7 ml can be injected i.v. into adult mice at each dose.

*Intraperitoneal injections.* The mouse is removed from the box by the tail with the right hand and lowered onto the grid work of a box lid. As the front feet grip the lid use the thumb and crooked outside edge of the index finger of the left hand to grasp the loose skin behind the head, pushing downwards firmly. The mouse can then be safely picked up and turned over. Tuck the tail beneath the clamped small finger in the palm so that the mouse is stretched, belly uppermost in the left hand. Swab the abdomen with alcohol. Using a 1 ml or 2 ml syringe and 10 mm, 25-gauge needle in the right hand insert the needle in about the centre of the lower left quadrant of the abdomen (right side of inverted mouse) at about 20° angle to skin and towards the chest. Depress the piston and withdraw. Volumes of up to 2 ml can be injected i.p. into an adult mouse at each dose. 1 ml is preferable as larger volumes may leak from the injection site.

*Intramuscular injections.* Up to 50 μl can be injected into each thigh muscle. The mouse is held as for i.p. injection. An assistant is then required to extend the hind leg. The fur is removed from the posterior aspect of the thigh with curved scissors, the site swabbed and injected using a 10 mm, 25-gauge needle.

*Subcutaneous injections.* These are best done from the i.p. position but injecting beneath the skin at the flexure of the dorsal thigh and abdomen; 0.2 ml can be injected on each side.

(ii) *The rat*

A similar approach can be adopted as for the mouse with all routes. Subcutaneous injections can be given to rats beneath the loose skin of the shoulder or flank by lifting a fold of skin and injecting up to 0.5 ml at a 45° angle through the skin. Intramuscular injections of up to 0.2 ml are given into the thigh muscles with a 25 mm, 20-gauge needle. This size is also used for i.p. injections, by which up to 5 ml can be administered. An experienced assistant is required to hold rats for immunization. For the i.v. route a restraining cage is required.

(iii) *The guinea pig*

*Intradermal injections.* These are easy to perform in guinea pigs (and rabbits). In both species the hair is carefully shaved and the site swabbed. A tuberculin glass syringe with 0.1 ml gradations and a 10 mm, 25-gauge needle are used. The needle point is gently inserted into the stretched skin at a close angle to the body and worked by gentle pressure and barrel rotation about 3–4 mm into the tissue. 0.1 ml is then carefully injected to cause a raised bleb in the skin. The needle is gently removed with further rotation and the site wiped to remove any returning fluid.

*Intravenous injections.* These can be performed using the front marginal ear vein. The barrel of a fine bore tuberculin needle 10 mm, 26-gauge is broken from its base and the blunt end fitted into one end of about 15 cm of nylon catheter tubing. The other end is attached to an intact needle, on a 1 ml or 2 ml glass syringe. The guinea pig is wrapped in a towel with the head protruding, the fur clipped with curved scissors from the front of the ear which is then warmed with cotton wool dipped in hot water. When the vein swells, the free needle is placed into the vein towards the head. It is advisable to use albino guinea pigs. Up to 2 ml can be injected into guinea pigs via the ear vein.

*Intramuscular i.p. and s.c. injections.* These are performed as for rats with similar volumes. Guinea pigs are acutely sensitive to mycobacterial proteins and s.c. injections of FCA may lead to open lesions. For this adjuvant minimum volumes should be used. The old practice of injecting FCA or FIA into the footpads is unnecessary to obtain a good antibody response. Small i.m. injections into the thigh from the dorsal aspect with the leg extended in the left palm is the advised approach.

(iv) *The rabbit*

*Intravenous injections.* These are performed with a 1 ml or 2 ml syringe and 16 mm, 25-gauge needle into the rear marginal ear vein. The surface of the ear is shaved with a scalpel blade and swabbed with 70% alcohol. The needle is introduced towards the head. The rabbit is best restrained on the lap of an assistant with gentle hand pressure on the shoulders. The usual dose is 1–2 ml but 5 ml can be given without difficulty.

*Subcutaneous injections.* Injections of up to 0.5 ml can be given through the loose skin of the shoulders or along the flanks using a 25 mm, 20-gauge needle. The site should be clipped of fur with curved scissors and swabbed. This is a common form of immunization with Freund's adjuvants as boosting injections.

*Intramuscular injections.* The rabbit is held on the lap and the operator extends one hind leg by gripping the knee from beneath and pulling the leg out firmly. The surface of the thigh above the knee is clipped of fur to expose the skin. Swabbing reveals the superficial blood vessels. A 40 mm, 21-gauge needle is used to inject up to 0.5 ml deeply into the thigh at a forward angle after ensuring that a deep blood vessel has not been penetrated. The needle should be withdrawn slowly and the site carefully swabbed to remove any returning fluid. This method is usually used for Freund's Adjuvant injections.

(v) *Sheep*

Sheep are easy animals to handle. For most procedures the seated assistant holds the animal in a sitting position from behind using the knees to support the back and the hands to tilt the head upwards and sideways. For i.m. injections in the upright position the animal is restrained from moving forwards by a firm arm across the chest.

*Subcutaneous injections.* These can be conveniently performed into the bare area of the skin at the flexure of the upper thigh/lower abdomen exposed in the sitting animal. The site is swabbed and skin raised by pinching. Immunogen in solution (1–5 ml) can be injected using a 25 mm, 20-gauge needle, or up to 0.5 ml of Freund's Adjuvant emulsion at each side.

*Intramuscular injections.* With the animal standing, the wool is clipped from a site at the back of the thigh and up to 1 ml of Freund's adjuvant emulsion is injected deeply into the thigh, using a 50 mm, 19-gauge needle and preferably a 1 ml glass syringe.

*Intravenous injections.* The neck is stretched by rotating the head sideways and tilting the chin upwards, flexing the neck over one knee from behind. The side of the neck is clipped free of wool. The thumb is pressed firmly to the side of the neck above the sternum to swell the external jugular vein. Swab the skin and inject into the vein using a 30 mm, 23-gauge needle on an appropriate sized syringe: 5 ml or more can be injected. Ensure the head is held firmly during injection.

## 5.2 **Taking trial bleeds**

*General point:* A preliminary bleed should always be taken from each animal prior to immunization, or in the case of mice a pooled sample from some animals will usually suffice.

### 5.2.1 *The mouse*

The mouse is restrained as for i.v. injections and the tail warmed in water. It is dried with tissue and gently rubbed with vaseline and the tip swabbed in 70% alcohol. A few mm of the tail are cut off with sharp sterile scissors; gentle massage down the tail produces successive drops of blood which are run down the side of a small glass tube. Up to 0.5 ml of blood can be collected once per week this way. The tip of the tail is compressed with an alcohol swab to complete the operation. With such small volume it is preferable to collect plasma and glass tubes coated inside with silicone Repelcote are helpful in preventing clotting and allowing the drops of blood to run to the bottom. A small volume (50 $\mu$l) of heparin/saline can be placed in the tube before bleeding. The blood should be centrifuged as soon as possible. For serum, small plastic precipitation tubes are used and the clot centrifuged to the bottom. Separate mouse samples are best stored in small conical capped tubes at −20°C.

### 5.2.2 *The rat*

Rats are usually bled by cardiac puncture under anaesthesia using an ether/air chamber (exhausted to the outside). The rat is removed on collapse and placed on its back on a cloth. A small beaker with ether-impregnated absorbent cottonwool in the base is placed in front of the nose to sustain anaesthesia. Using a 1 ml or 2 ml syringe, a 30 mm, 23-gauge needle is inserted into the chest cavity beneath the xiphisternum with the right hand, the left holding the chest cavity between index finger and thumb. With gentle withdrawal of the piston by a few mm to create a vacuum, insertion of the needle into the heart is easily detected. The syringe is slowly filled and the needle withdrawn. Rats recover rapidly from ether and can be placed back in the cage as soon as fully conscious. Up to 1−2 ml of blood can be harvested weekly from rats by this method. This is an expert operation and cannot be undertaken without adequate training.

### 5.2.3 *The guinea pig*

Guinea pigs are also bled by cardiac puncture but the appropriate anaesthetic is Fentanyl/Fluanisone (Hypnorm) by i.m. injection at the rate of 1 ml/kg body weight. Induction time in the guinea pig in our experience is at least 15 min by this route. Another method (31) describes the use of Hypnorm by the i.p. route for rats and this may also be used for guinea pigs. It is diluted 1 in 10 with distilled water and this given at the rate of 0.7 ml/100 g body weight.

### 5.2.4 *The rabbit*

Rabbits are bled from the rear marginal ear vein. The animal is placed on the lap and wrapped in a towel or surgical cloth to leave the head exposed outwards. The site is shaved. A very small amount of xylene may be applied to the tip of the ear with cotton wool to encourage vein dilation. The site is swabbed and rubbed above and below with vaseline. The tip of a pointed scalpel blade is used to cut across the vein and the vein compressed on the head side above the cut. The blood is collected into a glass bijou bottle, about 5 ml being the usual volume. An alcohol swab is clamped firmly to the cut with the fingers to stop the bleeding and during this time the xylene is washed completely from the tip of the ear with 70% alcohol and then water. After return to the cage the rabbit should be checked after a few minutes to ensure that the animal's cleaning has not opened the cut. Cyanoacrylate tissue adhesive is very useful for sealing the cut.

### 5.2.5 *Sheep*

Sheep are bled from the external jugular vein, the procedure being similar to i.v. injection. The needle (30 mm, 20-gauge) is inserted into the vein towards the head and 5−10 ml of blood withdrawn.

### 5.3 **Production and terminal bleeding**

#### 5.3.1 *Mice, rats and guinea pigs*

Exsanguination is performed following ether/air anaesthesia as described for trial bleeding of rats, or following Hypnorm administration in guinea pigs. The legs are fixed with tape to a board. For mice a 25 mm, 23-gauge needle with a 2 ml syringe is used; for rats and guinea pigs a 40 mm, 21-gauge needle with 10 ml syringe is used.

#### 5.3.2 *The rabbit*

For a succession of bleeds, rabbits can be bled up to 20 ml weekly or 50 ml every 2–4 weeks using the marginal ear vein or a butterfly needle with attached cannula inserted into the central ear artery. Rabbits can be sedated with Hypnorm using 0.5 ml/kg injected i.m. Terminal exsanguination is performed under anaesthesia by cardiac puncture. The rabbit may be given 2–2.5 mg/kg Hypnorm or an i.v. injection (ear vein) of pentobarbitone sodium (Sagatal, 60 mg/ml). For the latter a general guide is 0.5 ml/kg body weight but rabbits vary greatly in response—it is best to use a syringe holding 3 ml, inject the first 1 ml and then inject more, slowly, until the corneal reflex is lost.

Cardiac puncture follows the method for the rat using a 50 mm, 19-gauge needle and a series of 60 ml syringes or a transfusion needle and catheter leading to a sterile bottle. About 150 ml of blood can be harvested from an average-sized adult rabbit.

#### 5.3.3 *Sheep*

A good production bleed from a sheep is 500–700 ml of blood withdrawn from the external jugular vein using a human blood transfusion set. The collecting bag may contain anticoagulants, (i.e. acid citrate dextrose, ACD) for plasma collection or the blood may be allowed to clot in an anticoagulant-free bag and the serum poured off the following day. Bags with anticoagulant can be centrifuged at 4000 r.p.m. and the plasma likewise poured off. Sheep can be bled of the above volume on several occasions at monthly intervals under good nutritional conditions. However, when harvesting blood more frequently it is necessary to return the blood cells. Sheep are injected with 0.2 ml of heparin before bleeding. A total of 200 ml of sterile saline is added to the centrifuged cells in the ACD bag and the cell suspension is returned via the external jugular vein through a drip set over 5–10 min. The sheep is restrained in a cradle and can be mildly tranquillized with 20 mg Diazepam (Valium).

## 6. EXAMPLES OF IMMUNIZING REGIMENS

### 6.1 **Anti-hapten antibodies**

#### 6.1.1 *Anti-hapten response in mice and rats*

Alum-precipitated hapten-protein conjugate (i.e. DNP–MSH, oxazalone–MSH) is injected (mouse, 30 μg; rat 50 μg) with $10^9$ *B.pertussis* in a total of 0.5 ml of saline i.p. as a priming dose. The animals can be boosted at 21 days with 30–50 μg of immunogen alone in saline i.p. (mouse) or i.v. (rat). The animals are test bled 4 days post boost and thereafter. Booster doses can be given repeatedly i.p. With longer intervals between boosting injections (> 2 months) the doses can be reduced to 10 μg in saline. The maximal response is usually obtained about 6–7 days after each boost injection.

#### 6.1.2 *Anti-hapten response in guinea pigs*

Conjugate protein (50 μg) of, for example, TNP–BSA is emulsified in FCA and injected in the thigh muscle—total volume 0.2 ml per animal. After 3–4 weeks the animals are boosted with the same dose in FIA injected s.c. into several sites, each of 0.2 ml. The animals are test bled 2 weeks later.

#### 6.1.3 *Anti-aminoglycoside hapten response in rabbits (16)*

Conjugates of gentamicin and amikacin with PT and BSA are emulsified with FCA, using 1 mg of conjugate protein in a total volume of 0.5 ml per rabbit. This is injected into both thigh muscles. The rabbits are boosted with 100 μg of conjugate in FCA at 2-monthly intervals s.c.

#### 6.1.4 *Anti-glycocholic acid hapten response in rabbits (18)*

For weak immunogens it may be necessary to increase the dose of primary and boosting injections. Satisfactory antisera for the RIA of glycocholic acid were obtained by using 5 mg of hapten-carrier (BSA) conjugate in FCA in the first two (i.m.) injections and 100 μg of conjugate in saline for six i.d. injections over 4 weeks after 2 weeks' rest.

### 6.2 **Antibodies to single proteins**

#### 6.2.1 *In the mouse and rat*

A similar procedure can be adopted as for raising anti-hapten responses using protein–alum precipitate and *B.pertussis*. A recommended regimen (28) is to use $2 \times 10^9$ bacteria for the first i.p. injection in mice with 100 μg alum–protein in equal volume. A good antibody response to boosts of 10 μg immunogen alone are obtained if an interval of 2 months or more lapses between priming and boosting. Higher doses may be required for shorter time intervals.

#### 6.2.2 *In the rabbit*

For raising high titre precipitating antisera, of predominantly IgG isotype, to heterologous serum protein as an example, 100–200 μg of the protein is emulsified in FCA in 0.5 ml per rabbit and this injected into one thigh muscle. This is repeated in the other thigh after 10–14 days. A trial bleed may be taken about 10 days later; usually it is found to be necessary to give further boosting injections for high titre sera. A number of procedures can be adopted for this—all follow an ideal rest period of 2–3 months.

(i) Subcutaneous injections of protein in FIA in several sites (6–8 sites are recommended) the total dose each time being 20–100 μg.
(ii) A further injection of immunogen in FCA s.c. beneath the skin of the shoulder. We obtain excellent rabbit antisera to human immunoglobulins IgG, IgA and IgM by schedules (i) or (ii).
(iii) A series of i.d. injections of immunogen in saline, each of 10 μg in 0.1 ml.

(iv) Injections of 50–100 μg of protein in saline s.c.

Rabbits should be further test bled a week after boosting, or in the case of i.d. boosts, when a strong Arthus reaction develops at the injection site.

An excellent method for preparing anti-mouse immunoglobulin sera in rabbits is as follows.

(i) Immunize mice i.p. with weekly injections of $2 \times 10^9$ *B.pertussis* suspension for a course of 3–4 injections, repeated every 3 weeks.
(ii) Use the immune serum to coat a $10^{11}$ suspension of *B.pertussis*—0.5 ml serum/$10^{11}$ bacteria in 10 ml, incubate at 37°C for 1 h and wash in PBS.
(iii) Give each rabbit $4 \times 10^9$ antibody (IgG)-coated organisms i.v. for each of 4 weekly injections, and test bleed 1 week after the last injection.
(iv) The response is maintained by repeated courses after a rest of several weeks.

### 6.2.3 *In sheep*

Sheep respond well to i.m. or s.c. injections of 20–50 μg of purified heterologous proteins in FCA. Lambs of 3–4 months of age can be primed. A rest of several months is advised to gain the best response to boosting injections with 250–500 μg of protein, which are given also in FCA. Small volume s.c. injections give as good results as deep i.m. injections in the sheep without the formation of large granulomatous lesions. Trial bleeds are taken 1 week after boosts. This procedure raises high titre antisera to a wide range of heterologous serum proteins. For human IgG antisera the Fc fragment is purified as immunogen to reduce the requirement for absorption against intact other isotypes. For IgA- and IgM-specific antisera whole molecule is used as post absorption against IgG is simple [see Section 9.1.6 (ii)].

A similar immunization approach can be used to raise antisera to many antigens precipitated (i.e. with sheep antibodies) in agarose as the purifying principle. In these cases the amounts of heterologous protein injected are very much smaller (in the ng range)—yet as complexes and with repeated long interval boosting, good responses can be obtained.

Anti-immunoglobulins to species other than man can be prepared without prior immunoglobulin purification (see Section 3.2.3). For instance, to raise anti-rabbit IgM, 0.5 ml of the IgM fraction of decomplemented anti-SRBC serum was stirred for 10 min in 5 ml of 10% (v/v) SRBC. These were washed five times with PBS and the agglutinated cell pellet emulsified with CFA and injected i.m. After 3 weeks a second similar injection was given into three sites—a trial bleed 7 days later showed a good anti-IgM response by IEP against whole rabbit serum—absorption of anti-light chain antibodies with IgG rendered the antiserum specific to IgM. A similar approach has been adopted to raise IgG-specific antibodies using the IgG-rich eluted fraction of anti-red-cell sera. In this case contaminating anti-IgM antibodies were absorbed using a suspension of IgM-coated cells. This method appears to give bias to Fc isotype-specific epitope responses. The use of isotype-coated red cells for absorption is detailed later [Section 9.1.4 (iv)]. The approach is broadly similar to that used to raise anti-homologous immunoglobulin allotype antibodies in mice and rabbits (see Section 6.5).

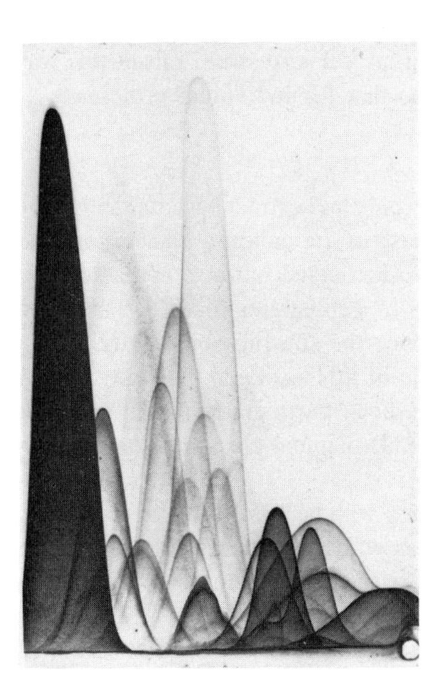

**Figure 7.** 2D-IEP of whole human serum showing the properties of a balanced multispecific antiserum to human serum proteins. (Print kindly donated by Dr A.R.Bradwell).

## 6.3 Antisera to complex protein mixtures

These antisera are raised to study components of antigen mixtures and to act as reference reagents in defining individual molecules. It is essential that both maximum diversity of response and a balance in titre of specificities is obtained so that on use at one concentration as many antigens of the mixture are displayed in, for instance, IEP or 2D-IEP. As starting material of, for example, a bacterial or parasite extract, cell membrane eluate or whole serum contains immunogens in widely varying concentration and intrinsic immunogenicity, a balanced antiserum that reveals minor immunogens is difficult to obtain—the tendency will be to gain a response to predominant large molecules. It may be necessary to fractionate mixtures and use many animals whose sera are ultimately mixed in varying proportions. Balanced polyspecific antisera require both patience and ingenuity.

### 6.3.1 *Anti-whole sera*

The major problem with whole sera or plasma as immunogens is to obtain a good response to molecules in the $\alpha$ globulin region. The best approach is to separate serum by agarose gel horizontal slab electrophoresis. Strips of the run gel cut across the migration path can be frozen and the components harvested on thawing. Sheep can then be immunized with different component fractions according to electrophoretic mobility, each assessed by IEP. Antisera to individual components of sera—for example complement components, can be added. *Figure 7* shows an example of a well-balanced antiserum to human serum proteins in sheep raised this way. The general method of immunization is similar to that for individual proteins.

### 6.3.2 *Bacterial antigens*

There are many ways to prepare and extract bacterial antigens (32). These and the growth conditions and growth phase of the cultured bacteria will determine the nature of the product for immunization. Sonicated extracts of bacterial cell walls are enormously complex mixtures, as *Figure 4* illustrates for *E.coli* and mycobacteria, and numerous ways are available to reduce the starting complexity by physical and chemical fractionation. A good example of this has been the study of mycobacterial antigens (33). Polyspecific antisera have been particularly useful in further defining the molecular components of Seibert's (34) original classification of the major polysaccharide and protein fractions.

In our hands immunization with heat-killed or irradiated or ground powder of *M.bovis* BCG or *M.tuberculosis*, as with other non-solubilized bacterial preparations, in FIA in rabbits leads to a response restricted to dominant surface molecules—mainly proteins in the case of mycobacteria. We have used $5 \times 10^6$ intact bacilli or 1 mg of ground powder in FIA i.m. followed by repeated s.c. injections at long intervals of 3–6 months to raise such antisera in rabbits. For a broader range of specificities it is necessary to use an immunogen solution. There is no standard mycobacterial soluble antigen mixture for immunization. Partial fractionation can be achieved by many methods to yield major protein and carbohydrate moieties and glycolipids (32). We use the same method for membrane antigen extraction as for *E.coli*—that is by sonication, detergent and centrifugation (see Section 3.3.2). 100 μg of protein is used in FIA repeatedly. Antisera to the purified protein derivative (PPD) of *M.tuberculosis* are prepared in the same manner in rabbits. The major proteins of PPD are antigen 6 and antigen 7 in the Janicki (US–Japan) nomenclature (32).

Mice respond well to PPD. High titre ELISA positive sera (indicating a suitable response for monoclonal fusions) was obtained by injecting 50 μg of PPD in FIA i.p., boosting on days 30 and 60 with 500 μg in FIA i.m. and giving 1 mg i.p. in saline 4 days before fusion. A similar regimen was used to raise antibodies to *M.tuberculosis* and *M.bovis* BCG sonicate antigens, using 200 μg in FIA as the priming dose.

A useful means of separating bacterial proteins and carbohydrate for immunization, which we have applied to mycobacterial sonicates, is to utilize the fact that carbohydrates spontaneously bind to SRBC, whereas cells treated with tannic acid adsorb only proteins. Washed 3% (v/v) red cells (10 ml) are incubated at 37°C with 1:100 (final dilution) sonicate antigen for 2 h, centrifuged and washed repeatedly. The tannic acid method was as described originally by Boyden (35).

(i) Add 0.25 mg of tannic acid to 10 ml of 0.2 M PBS, pH 7.2 and mix this with 10 ml of 3% (v/v) SRBC suspension. Incubate for 10 min at 37°C.
(ii) Wash the cells once with saline, centrifuge and resuspend the pellet in 0.5 ml of sonicated BCG supernatant.
(iii) Incubate this at 37°C for 2 h, then add 2 ml of Hepes medium, centrifuge and wash again.

(iv) Finally resuspend the cells in 4 ml of Hepes with 0.1% BSA.
(v) Immunize mice repeatedly with 1 ml of a three-times diluted suspension i.p. of carbohydrate- or protein-coated red cells.

6.3.3 *Parasite extracts*

The essential principle for raising precipitating multispecific antisera to helminth parasite surface antigens or total extracts, of high titre, is repeated low-dose immunization in FCA with lots of patience. Antigen extraction methods will vary according to parasite (36). For *Schistosoma* species, surface antigens of cercariae, 5-day-old lung form schistosomula and young and mature adults can be eluted by slow rotary mixing in 4 M KCl overnight at room temperature. After settling, the supernatant is centrifuged at 10 000 r.p.m. at 4°C for 1 h and dialysed against PBS. For total extraction the parasites are subjected (in 4 ml volume of PBS) to three 5 min sonication cycles in an ice bath with intervening 5 min cooling periods using 0.8 Amp output from a 60 Watt ultrasonic probe. The sonicate is centrifuged at 10 000 r.p.m. and the supernatant used as immunogen.

In rabbits and sheep 50−100 μg of protein is injected in FIA i.m. and a second i.m. injection given 2 weeks later. Boosts with 50 μg are given in FCA s.c. at 3−6 monthly intervals with a trial bleed 2 weeks after each boost. *Figure 2* shows the broad specificity of one such antiserum to *S.mansoni* (*Figure 2a*) and the value of using excised precipitation peaks from plates such as these to produce unispecific antisera (*Figure 2b*).

6.4 **Antisera to bacterial suspensions**

6.4.1 *In mice*

(i) *To M.bovis BCG.* High titre ELISA-positive antisera to BCG coat proteins is raised by injecting $2 \times 10^3$ organisms of BCG vaccine suspension in FIA i.p., followed at 30 day intervals with $10^4$ vaccine in saline i.p. rising to $10^6$. For subsequent monoclonal spleen cell fusions the last injection of $10^6$ is 4 days prior to fusion.

(ii) *To B.pertussis*. Mice are injected weekly i.p. with $2 \times 10^9$ *B.pertussis* suspension in saline in 0.2 ml volume. Successive 4 weekly courses, with intervening rests of 3 weeks can be given. Anti-pertussis antibody production, with a predominant $IgG_{2a}$ isotype ($\alpha$2a) response, is a prelude to the preparation of between-strain anti-mouse immunoglobulin allotype antibodies, especially those relating to the *Igh*-1 locus of $\gamma$2a molecules (see anti-allotype production, Section 6.5.1). Mouse antibody-coated pertussis can also be used to raise anti-mouse immunoglobulin sera in heterologous species (see Section 6.2.2).

6.4.2 *In rabbits*

Rabbits are an obvious choice for i.v. immunization with killed bacteria. One major exception is mycobacterial suspensions which induce highly pathogenic inflammatory lesions.

Many successful antisera to Gram-positive and Gram-negative bacteria have been produced. The tendency for the response to predominate against serotype/strain-delineating carbohydrate structures of the LPS of Gram-negative organisms has meant that rabbit antisera can be used as valuable serotyping reagents. Cross absorption with cultures of other types and species is easy to perform and removes antibodies to common outer membrane protein epitopes.

The general principle for bacterial suspension immunization is to use a rising dose of organisms in once- or twice-weekly i.v. injections in courses of 3–4 weeks, resting the animal for several weeks between courses. The scale usually begins with $10^6-10^7$ organisms and rises to $10^9$ or greater as the response gets under way. This compensates for antibody *in vivo* bacterial neutralization and rapid clearance of opsonized organisms from the lymphoid tissues.

(i) *Campylobacter jejuni serotypes 1, 2 and 4 (37).* The bacterial cultures are harvested in log phase at about $10^8$/ml. The washed and formalized suspension is first diluted to give an OD reading of 0.375 at 625 nm. In our hands a first course of 1 ml i.v. twice weekly for 3 weeks, with subsequent further courses of 2 ml and 3 ml on each injection yields, 1 week after the final injection, very high titre antisera. These can be absorbed against other Gram-negative organisms and other Penner serotypes (37) to yield highly specific reagents.

(ii) *Type III pneumococci – anti S3 carbohydrate.* Paradoxically the purified S3 carbohydrate of pneumococci is not, or is poorly, immunogenic in rabbits. However, in association with membrane protein on heat-killed organisms it induces a phenomenal IgG antibody response by chronic i.v. injection which is highly carbohydrate- and type-specific. With repeated courses rabbits produce at least 5 mg/ml of specific antibody and many animals may respond by elevating their serum IgG antibody concentration to over 20 mg/ml. The response is clearly not influenced by the normal homeostatic feedback control on antibody synthesis exercised with protein immunogens. In one experiment (38) our rabbits were given at least two courses of nine i.v. injections over a period of 19 days beginning with $10^8$ organisms rising by the sixth injection to $4 \times 10^9$ organisms and totalling $10^{10}$ organisms. Rabbits were bled 3 or 4 days after the last injection.

(iii) *Proteus vulgaris* ($\phi$X19). High titre antibodies of IgG isotype are raised to *Proteus* carbohydrate using formalized suspensions. Raising such antisera in immunoglobulin-allotyped rabbits is a first step in preparing homologous (rabbit anti-rabbit) immunoglobulin allotype-specific reagents, using a single allotype mismatch of antibody donor to recipient as an antibody-*Proteus*-complex.

Rabbits are given as a first course an injection of $10^9$ organisms, followed by two injections of $2 \times 10^9$ on days 8 and 11 and $5 \times 10^9$ on days 15 and 18. After 3 weeks a similar course is repeated. Rabbits are bled 3 days after each course—the titre remains high for up to 10 days and two more bleeds may be taken.

### 6.5 Anti-immunoglobulin allotype sera

#### 6.5.1 *In the mouse*

An excess of anti-*B.pertussis* antibody of an inbred strain (e.g. C57BL) which carries an immunoglobulin allotypic allele (heavy chain structural variant for the type strain) is used to coat a suspension of $10^{11}$ *B.pertussis* for 2 h at 37°C. After washing,

$2 \times 10^9$ antibody-coated organisms are injected i.v. (or i.p.) into the recipient, allotypically different mice (e.g. BALB/c strain). A course of 4 weekly injections is given which may be repeated after 3 weeks. Further information on this system has been recently reviewed (39).

### 6.5.2. *In the rabbit*

Rabbits are immunized with suspensions of antibody-coated *Proteus* in courses identical to that inducing the anti-*Proteus* response in the donor. The number of courses required depends on the immunogenicity of the allotypic allele under immunization (40).

## 6.6 **Anti-immunoglobulin idiotype responses**

### 6.6.1 *Principle*

The antigen binding site regions of the antibodies are structurally complex to the extent that each antibody specificity (clonal product) presents a set of epitopes (idiotopes) to the immune system, collectively called the idiotype, which can stimulate an autologous anti-idiotype antibody response. Perceived as a set of internal recognition sites for expansion and regulation of antigen-responsive cells, the network of idiotypes and autologous anti-idiotypes is usually an unmeasured, infinitely complex entity of serum and cells. However, conditions of extreme provocation such as experimental return injection of an individual's purified antibody in adjuvant, will induce a distortion of the network and a large autologous anti-idiotypic response. The implications of this lie not only in formal demonstration of a strand of the network and the proof of idiotypes, but also in confirmation of a major prediction of network theory that some anti-idiotypes have idiotopic structures that are 'internal images' of epitopes of the first (exogenous) inducing immunogen. Thus anti-idiotypes can be prepared that act as surrogate immunogens both as an immunizing principle (41) and as target antigens in immunoassays. The intrinsic immunogenicity of purified antibody, via idiotypy, results in potent anti-idiotype responses after cross-immunization between animals of the same species [isologous or homologous (allogeneic) anti-idiotype] and between different species (heterologous anti-idiotype). The most success in inducing anti-idiotype antibodies with surrogate antigen/immunogen properties has come from the latter two approaches. Examples of some methods to produce anti-idiotypes are given below.

### 6.6.2 *Mouse anti-mouse idiotype (isologous)*

High titre isologous anti-idiotypic antibodies have been raised to Mabs in BALB/c mice which were immunized with immune complexes of a purified IgG human paraprotein ($IgG_1$ subclass) bound with individual BALB/c Mabs (42). These were towards an epitope of the $\lambda$ chain or to constant heavy chain epitopes of the $C\gamma2$ or $C\gamma3$ domains. Other successful complexes contained a pool of Mabs to the human $\gamma$ chain or $\lambda$ chain. Immune complex precipitate for immunization was prepared by adding Mab (or mixtures) (centrifuged ascitic fluid), in slight excess of optimal proportions, to 1 mg/ml of the human paraprotein in the presence of 5% PEG (3000 mol. wt). The precipitates were washed in the cold with 5% PEG and taken up in PBS to 1 mg/ml paraprotein. Mice were primed by i.p. injection with 100 $\mu$g of paraprotein as complex, mixed with

$10^9$ *B.pertussis* and Alhydrogel. Mice were boosted i.p. with the same amount of complex without *B.pertussis* on days 6, 13, 20 and 27, or on days 26, 40 and 48. These schedules yielded high titres of isologous anti-idiotypic antibodies (7 days after the last injection) specific to the complexing monoclonal(s) as well as separate antibodies to the human paraprotein.

### 6.6.3 *Rabbit anti-rabbit idiotype (homologous)*

Rabbits have immunoglobulin allotypic determinants coded at several genetic loci. Being outbred it is not possible to avoid these differences by using the same inbred strain for donor and recipient and in consequence the animals require matching as closely as possible, at least at the major *a* and *b* loci. Anti-idiotype antibodies are easy to raise to the idiotypes of the anti-*P. vulgaris* carbohydrate response, in allotype-matched animals. Similar courses of immunization with antibody-coated bacteria are used as those to generate anti-allotype; however, four or five courses may be necessary.

It is also possible to raise anti-idiotype responses to affinity-purified rabbit antibodies without complexing the combining sites (43). This is achieved by the simple procedure of using 100–200 μg of antibody in FIA (equal volumes) as repeated small i.m. injections at intervals of several months.

A more rapid response can be obtained by repeated i.v. injection of purified rabbit antibody coupled to the recipient rabbit's own red cells by the glutaraldehyde coupling technique (43).

(i) Harvest fresh red cells in heparin and wash in sterile saline.
(ii) Add packed cells (15 μl) suspended in 300 μl of rabbit antibody (300 μg) in 0.15 M phosphate buffer and 20 μl of 2.5% glutaraldehyde solution.
(iii) Mix and leave for 1 h at 4°C, then wash the cells three times in sterile saline and resuspend in 3 ml.

Rabbits are injected with 1 ml of coated cells on days 1, 7, 10, 13 and 17. The course is repeated after 2 weeks and the animals bled 1 week after the last injection. Anti-idiotype specificity can be sensitively tested using the same method for antibody (idiotype) coupling to cells and performing haemagglutination reactions with a panel of different idiotype antigens.

### 6.6.4 *Heterologous rabbit anti-mouse idiotype (44)*

Rabbits were immunized with the globulin fraction of mouse monoclonal ascites fluid obtained by precipitation with 18% sodium sulphate and dialysis against PBS. Protein (2.5 μg) was used in each injection, the first given i.m. in both hind legs in FIA. Booster injections were given after six weeks at two s.c. sites in FIA and in the fourth month by this route without adjuvant. The rabbits were bled 10 days later.

In heterologous immunizations a substantial part of the antibody response is to isotypic and allotypic determinants on the foreign species antibody. This requires extensive absorption. In the above case successful absorption was achieved by passing the rabbit antisera (5 ml) repeatedly through a 10 ml column of normal mouse immunoglobulin bound to a Sepharose gel, until all trace was removed of antibodies to normal mouse immunoglobulin and to mouse monoclonals other than the donor source.

### 6.6.5 *Guinea pig anti-rabbit idiotype (heterologous)*

A useful approach to specific production of heterologous anti-idiotype has been to suppress the unwanted response to donor species-related isotypic determinants of the antibody (45). Guinea pigs were injected i.v. with 5 mg of de-aggregated (ultracentrifuged) normal rabbit IgG just prior to injection in FCA of 100 $\mu$g of affinity-purified homogeneous anti-streptococcal carbohydrate IgG antibody. The suppression routine resulted in a predominant response to the idiotype of the immunizing heterologous anti-streptococcal antibody.

## 6.7 **Antisera to sheep red blood cells**

Immunizing animals with sheep red blood cells (SRBC) is instructive in demonstrating the kinetics and cellular basis of IgM and IgG antibody responses and in gaining the practical skills of red cell agglutination and haemolytic titrations and haemolytic plaque assays. In addition the coating of red cells with anti-erythrocyte antibodies of defined isotype, allotype etc. forms the principle of antiglobulin haemagglutination reactions as one method for the specificity testing and titration of antiglobulin reagents.

Sheep red cells are harvested from blood withdrawn under sterile conditions into anticoagulant (ACD). Aliquots of whole blood (20 ml) are centrifuged at 1500 r.p.m. for 20 min in sterile universals and the plasma and buffy coat removed. The cells are then washed four times in 20 ml Alsever solution and resuspended in this volume. They can be stored for a week at 4°C in this condition. For immunization the cells are washed twice in PBS and are made up to a 10% (v/v) suspension in PBS which contains approximately $10^9$ cells/ml.

### 6.7.1 *Mice*

Animals are injected i.v. or i.p. with approximately $10^8$ cells (0.5 ml of 2% cells). The antibody response 5–6 days after a single injection is predominantly IgM. Repeated injections twice per week for up to 4 weeks will induce a predominant IgG response. Ideally the start of a course after a priming injection should be delayed for 2 weeks until the first IgG response is diminished. The doses may be increased to 0.5 ml of 10% cells.

### 6.7.2 *Rabbits*

A first i.v. injection of $4 \times 10^9$ cells (or 1 ml of 10% cells/kg body wt) will, as in mice, induce an IgM antibody response after 5–6 days. They can then be given an intensive course of about ten further injections, daily, for the first four and then on alternate days. Repeated courses spaced by several months can give high IgG agglutinating titres. Patience is required to obtain the best antisera.

## 6.8 **Antisera to mammalian cell membrane antigens**

The difficulties in raising polyclonal antisera to intact nucleated cells, or cell membrane extracts, by intra- or inter-species immunization, that are valuable in defining discrete cell surface antigens has long been recognized. Except in the special circumstances of molecules of the major histocompatibility complex and some other notable antigens of lymphocytes that are polymorphic or allelic systems, the absorption of the induced polyspecific sera is too problematical usually to give acceptable results by this approach. Examples of where it has proved useful in the mouse have been reviewed (28).

The reason why this important area is not addressed more fully here is that conventional polyclonal methods for raising anti-cell-surface antibodies have been superseded by the Mab approach which is considered in this context in Chapter 3.

## 7. PREPARATION AND STORAGE OF SERA

### 7.1 **Harvesting serum**

Blood should be allowed to clot for 1 h at 37°C. At this stage the clot should be freed from the walls of the container— for glass universals we use a sterile defibrinating stick— and the blood is best left at 4°C overnight for the clot to retract. Serum with the minimum of red cells can be withdrawn by pipette to tubes or universals with a conical base. The serum is spun at 1500 r.p.m. for 20 min to remove red cells. Preservative, when used, is added at this stage. Serum samples should not be pooled before testing (with the possible exception of mouse trial bleeds), but stored separately. Untested sera can be stored as individual whole bleeds but with the required volume for testing withdrawn as an aliquot. This avoids thawing the bulk. We find it convenient to tape the test aliquot in a capped tube to the side of each bulk container.

### 7.2 **Storage containers**

Bulk containers may be screw-capped polypropylene bottles of 50–500 ml volume (do not use glass as these may crack on cold storage) or small universals or capped tubes. An airtight seal is essential for all containers so the quality of container is important. Rubber cap liners and stoppers deteriorate on cold storage and should not be used. Containers should be labelled with the animal number and date of bleed—consecutive bleeds (and the pre-bleed sample) stored as a group. Do not write directly on glass bottles as this may come off on thawing. Use a complete ring of masking tape and an indelible pen. Small volume samples can be stored in plastic envelopes with a label inserted and the tubes numbered by a code referred to on the label. Aluminium trays of about 20 cm width with deep sides that slide the depth of the refrigerator are ideal for serum storage. For convenience deep freezers should be of upright design with half doors, each half containing three or four fronted shelves. Catalogues referring to unit, shelf and tray number allow easy location of samples.

### 7.3 **Preservatives**

Thiomersal (Merthiolate) 1:10 000 or 0.1% (w/v) sodium azide are recommended for long term storage of sera. Neither should however, be used for sera applied to animals. Sodium azide inhibits chemical coupling reactions and should be removed by prolonged dialysis. It is also toxic and should be handled carefully. Solutions should be stored in a locked cupboard. Preservatives are unnecessary with sterile sera or sera stored at −70°C.

### 7.4 Freeze storage

Although sterile antisera remain potent when stored at 4°C for long periods, deep freezing is generally recommended with a minimum of −20°C. To avoid repeated thawing, samples should be stored in minimum practicable volumes—for valuable reference reagents these might be 100 μl aliquots in capped tubes. At −20°C on long storage the water component sublimes and re-freezes as ice at the top. Concentrated proteins will aggregate to insoluble precipitate at the bottom. Thawed samples should be thoroughly mixed and centrifuged at high speed before aliquoting. Enzymes in sera continue to degrade proteins stored at −20°C. Storage at −60°C or below is ideal but an expensive luxury for all but the most valuable reagents.

### 7.5 Freeze drying

Sera vary in solubility after lyophilization. This is best reserved for small samples of valuable reagents pre-tested for solubility and activity after reconstitution, for samples posted over long distances and where refrigeration is unavailable. Mark lyophilization vials with the pre-dried level as a guide for reconstitution with distilled water.

### 7.6 Plasma

Harvesting of plasma affords a greater volume with less haemolysis. Bulk plasma from sheep and other large animals is best obtained by using anticoagulant blood transfusion bags which can be filled under sterile conditions. For small volumes, citrate (3.8%), EDTA (2%) is added to the syringe or bottle as 10% (v/v). Glass syringes rinsed in anticoagulant are used—these and the vessels should be cold (on ice). The blood should be centrifuged immediately at 4°C with a pre-cooled rotor, at 2500 r.p.m. for 20 min. To prepare serum from plasma add 2 ml of 10% anhydrous $CaCl_2$ per 10 ml citrate, or 1.6 ml of 2% $CaCl_2$ per 10 ml EDTA with stirring. Maintain at 4°C for several hours. The formed clot requires compression to withdraw the maximum serum. Freezing and thawing may be used to collapse the clot which can then be centrifuged or filtered out.

## 8. PRELIMINARY TESTS ON ANTISERA

Production of the best antisera depends on a strategy of trial bleeds after each injection to monitor the rise in antibody titre and changes in specificity and avidity. Test results will influence the nature (i.e. adjuvant, dose, route) and timing of subsequent injections and will be an essential guide to selecting the best responding animals and the absorption required in the final harvest.

The golden rule in assessing trial bleeds is always to include the test system for which the antiserum is ultimately intended, as an antiserum may seem adequate by, for example, GDD analysis but have poor specificity and/or low avidity in the RIA for which it has been raised.

A range of tests useful for assessing trial bleeds has been provided in Table 2 of Chapter 1. Different tests for titre and specificity are applicable according to the nature of the test antigen and the antiserum's intended use. *Table 5* sets out in more detail

**Table 5.** Test systems for antiserum trial bleeds and quality control.

| *Antiserum to:* | *Titre* | *Specificity* |
|---|---|---|
| *Small molecules* (raised as anti-hapten) | PHA: red cells coupled with hapten or with hapten linked to second carrier. ELISA/IRMA: plates coated with hapten linked to second carrier. Developed with labelled antiglobulin. EMIT: uses enzyme-coupled small hapten. RIA: uses radiolabelled hapten or conjugate (Volume II). Equilibrium dialysis: affinity assay for homogeneous response to small chemical haptens (see ref. 51). | GDD: antiserum versus (i) homologous hapten-carrier, (ii) hapten on second carrier, (iii) different hapten on homologous carrier linked by same chemistry, (iv) carrier alone (Chapter 6). ELISA/IRMA: with (i) to (iv) above as coating antigens (Volume II).. RIA: uses labelled standard antigen: compare binding curve of test and reference antisera (Volume II). |
| *Purified single macromolecules* | GDD: dilutions of antiserum in ring of wells round central antigen well. RRID: uses antigen-containing agarose gel and antisera in wells. Compare with reference antiserum (see *Figure 8*). PHA: uses antigen coupled red cells. ELISA/IRMA: uses antigen-coated plates (see *Figure 9*). RIA: uses radiolabelled antigen. Quantitative precipitin test: absolute concentration of antibody determined from optimal proportion precipitate (see Volume II). | GDD: antiserum versus immunogen, standard antigen, original mixed antigen source, related proteins etc. IEP: antiserum versus antigens as for GDD above (see *Figure 10*). 2D-IEP: uses antiserum in agarose and original mixed antigen source to look for single precipitation peak (*Figure 2b*). PHA, ELISA/IRMA: uses panel of antigen-coated red cells or wells (see *Figure 9*). WB: use SDS–PAGE tracks of original mixed antigen source, purified antigen standard and related proteins etc. |
| *Immunoglobulin isotypes* | Tests as for other single macromolecules: for antigen isotype standards in man use purified immunoglobulin paraproteins (myeloma proteins), polyclonal isotypes of normal sera, urinary Bence Jones light chains ($\kappa$ and $\lambda$) and IgG Fc($\gamma$) subunits. Mouse standard isotypes are purified Mabs (non-antiglobulin). Myeloma proteins are available in the rat. For other species prepare isotypes from pooled normal sera. | |
| *Antigen mixtures* | Titre not as relevant as balanced multispecificity. IEP: antisera in troughs should produce good precipitation arcs at one filling. 2D-IEP: antisera should be $<10\%$ (v/v) of agarose. Look for balance of peak heights (see *Figure 7* and Chapter 6). | IEP: look for number of precipitation arcs against standard antigen mixture. Compare with reference antiserum (see *Figure 10*). 2D-IEP: as for IEP in relation to precipitation peaks (*Figure 7*). Western blotting: compare binding pattern to standard antigen with reference antiserum. Use mol. wt markers to define antibody specificities (see Chapter 6). |
| *Bacterial extracts* | PHA: Protein antigens coupled to red cells by tannic acid or chromic chloride. Carbohydrates passively adsorbed. IEP/2D-IEP: as for antigen mixtures. ELISA/IRMA: for overall titre to various bacterial extracts. Also for titre to purified antigens (Volume II). RRID: use purified single antigens (i.e. LPS, polysaccharides etc.) in agarose. Compare with reference anti-LPS etc. | IEP/2D-IEP: as for antigen mixtures, ELISA/IRMA: antiserum versus (i) immunogen; (ii) other extracts of immunizing species; (iii) antigens of other strains, serotypes and species. Western blotting: very useful for determining cross-reactive antigens of other bacteria for absorption (see Chapter 6). |

| | | |
|---|---|---|
| *Bacterial suspensions* | Bacterial agglutination dilution assay IFT: dilution assay with labelled antiglobulin. | Bacterial agglutination and IFT: Use intact cells of same and related strains, species etc. |
| | Other tests as for extract antigens with emphasis on separated surface molecules. | |
| *Red cells* | Direct haemagglutination titrations. Haemolytic titrations with complement: combined IgM and IgG titre with untreated sera, IgG titre after 2-mercapto-ethanol treatment (Chapter 7). | For SRBC not usually applicable but other species red cells can be used. Specialized red cell antigen typing in man (see Volume II). |
| *Other cells in suspension* | IFT/IPT: direct or with labelled antiglobulin (Volume II). Complement-dependent cytotoxicity with exclusion dye or $^{51}$Cr-release assay. | IFT: greatly facilitated by FACS analysis; antiserum tested for capacity to inhibit binding of labelled reference conjugate (Volume II). |
| *Antigen in tissues* | Specificity for tissue sections determined by (i) testing on standard antigen-positive and negative tissues or cells; (ii) absorbing out the desired specificity with pure antigen or positive cells before application and (iii) blocking specific binding by section pre-treatment with unlabelled reference antiserum before application of labelled test reagent (Volume II). | |
| *Immunoglobulin allotypes* | Confirmation of allotypy depends on demonstration of Mendelian inheritance of alleles at allotype loci expressed as co-dominant antigenic markers. New typing sera require comparison with reference reagents against standard antigens. Some locus alleles in the rabbit can be typed with precipitating antisera by GDD. Others require specialized PHA or RIA tests. In the mouse, alleles of different isotype loci are expressed in different inbred (type) strains. They are determined by RIA (see refs 52 and 53). | |
| *Immunoglobulin idiotypes* | Proof of idiotypic specificity lies in the restricted binding to the inducing antibody molecule (idiotype) and not to other antibodies of the same or other animals. The usual source of idiotype immunogen (and test antigen) is a homogeneous, affinity-purified (anti-carbohydrate) antibody, myeloma protein or (mouse, rat, human) Mab. Idiotype–anti-idiotype reactions (and titres) are tested by GDD, PHA (idiotype-coated red cells), ELISA/IRMA or RIA. | |

some relevant test systems by antigen category as a guide to test procedure. Working protocols for all of these are to be found in Chapters 6 and onwards of this book and in Volume II.

## 9. ABSORPTION OF ANTISERA

### 9.1 **Antisera to single antigens in solution**

Most antisera for single antigen studies require removal of unwanted specificities. This is unavoidable using even the purest immunogen if it shares some epitopes with related molecules present in the test samples to be assayed. This is a common problem in hormone assays. In addition there are usually some antibodies to contaminating molecules of the inoculum which may also be present in the test samples. Affinity purification on an antigen column may be advantageous for other reasons (e.g. antibody concentration, antibody labelling efficiency, reduced background) but for specificity, for the reasons given above, it may not solve the problem unless an epitope-restricted, antigen-specific subunit or peptide can be prepared. For most purposes absorption of antisera is a more practicable solution once cross-reactive antigens and contaminants have been defined. A major advantage of absorption (as used here to mean the removal of un-

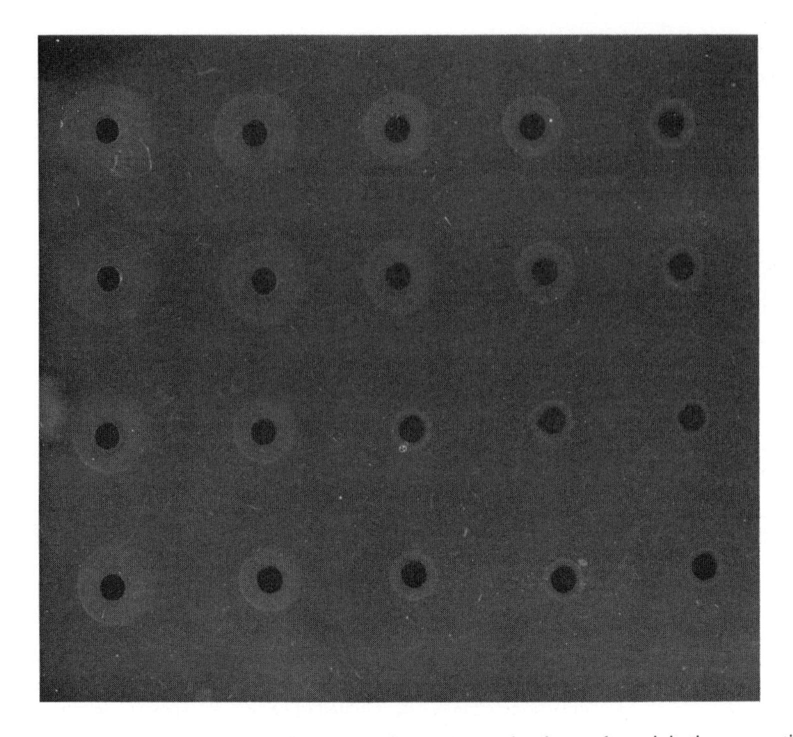

**Figure 8.** Reverse radial immunodiffusion test used to compare the titre and precipitating properties of a newly-produced antiserum to human IgG (**lower two rows**) with a reference antiserum (**upper two rows**). By comparison of precipitation ring sizes the test antiserum was shown to contain 50% less antibody than the reference reagent. The agarose contains 10 $\mu$l of whole normal human serum in 9.8 ml 1% agarose containing 3% (w/v) polyethylene glycol. The sera are diluted (left to right in two rows) 1:2, 1:4, 1:8, 1:16, 1:32. Wells received 5 $\mu$l and the plate has been kept at room temperature for 24 h.

wanted antibodies) is that the unbound harvest of antibody is not subjected to the harsh denaturing conditions required for dissociation of bound antibody in affinity purification methods. Furthermore the recovery of antibody tends to be poor in the latter approach and usually only the lowest affinity (most easily dissociated) antibodies can be harvested in active form.

Absorptions should be performed on a solid phase and this requires the chemical coupling (or passive binding) of adsorbing antigens to an insoluble support (e.g. derivatized cellulose, polystyrene, red cells etc.) or the preparation of a cross-linked antigen gel. The preparation and use of some commonly used immunoadsorbents is given below.

### 9.1.1 *Cross-linking of adsorbing proteins with glutaraldehyde*

(i) Prepare 250 mg of protein in 5 ml of phosphate buffer, pH 7. This can be wholly adsorbent antigen, or if less is available it can be mixed with serum albumin. Dialyse against the phosphate buffer and check the pH.

(ii) Add 1−2 ml of 2.5% aqueous glutaraldehyde dropwise with stirring until a gel begins to form, adjusting the pH with NaOH to neutrality where necessary.

(iii) Allow the gel to stand for 3 h at room temperature and then mash to fine particles.

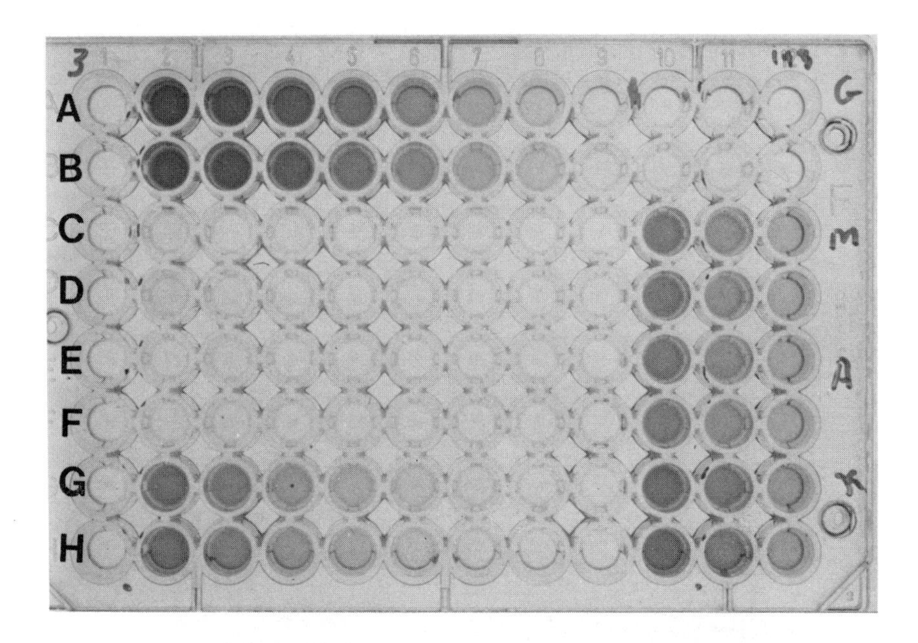


**Figure 9.** An ELISA test for the specificity of an antiserum to human IgG. The plate is coated with antigens (2.5 $\mu$g/ml) as follows: **rows A** and **B** with human IgG, **C** and **D** with human IgM, **E** and **F** with human IgA (all prepared from normal serum) and **G** and **H** with $\kappa$ light chain. The test antiserum was applied in doubling-dilutions (starting at 1:100) from **column 2** across the plate to **column 12** (**rows A** and **B**) or **column 9** (**other rows**).; **Columns 10–12** of **rows C–H** received dilutions of antisera specific to the coating antigens of these rows. An enzyme (HRP)-conjugate of rabbit anti-sheep IgG was used to develop the plate. **Column 1** is a conjugate alone control. The test antiserum reacts only with human IgG. The reaction in **rows G** and **H** is due to an IgG contamination discovered in the $\kappa$ chain preparation used for coating. Thus the test also has been used to demonstrate the purity of coating antigens.

(iv) Wash the gel copiously with water, and then with 0.2 M phosphate buffer, pH 7.4, 0.1 M glycine–HCl buffer, pH 3.2 and then repeatedly with PBS, pH 7.4.

(v) Quench any residual reactive sites with 1% (w/v) BSA in PBS (equal volume to gel) for 2 h at room temperature or overnight at 4°C. Wash again as in (iv) above.

(vi) For absorption of undiluted sera mix with adsorbent polymer in equal volume and stir (a rotary mixer is best) at room temperature for several hours. The absorbed sera can then be recovered by centrifugation.

(vii) The immunoadsorbent can be cleared of bound antibody and used repeatedly. For reconstitution wash copiously with 0.1 M glycine–HCl buffer at pH 3.2 to dissociate antibody, and then with PBS until the pH of the eluate rises to 7.2–7.4 and the optical density at 280 nm is minimal against a PBS blank. Washed adsorbent can be stored in a PBS suspension with 0.1% (w/v) or sodium azide at 4°C.

### 9.1.2 *Sepharose–antigen columns*

Sepharose 4B (Pharmacia) can be activated to a protein-coupling condition with CNBr. The activated form can also be purchased (Pharmacia) and used for step (iv) onwards.

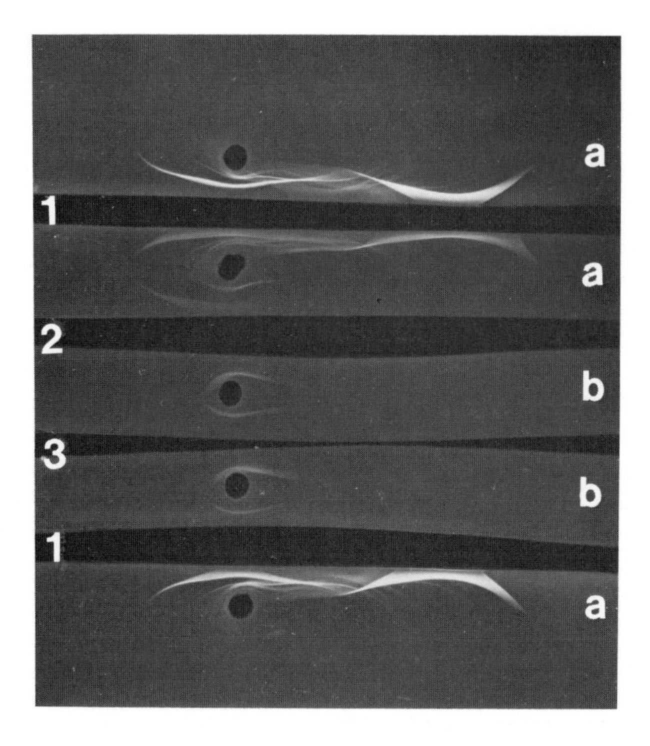


**Figure 10.** Immunoelectrophoresis used to test the specificity of an antiserum raised to human IgM before and after absorption with other immunoglobulin classes. Wells: (**a**) contains normal human serum, (**b**) contains purified IgM. Troughs: (**1**) contains anti-whole human serum, (**2**) contains unabsorbed test antiserum, (**3**) contains the test antiserum after absorption with IgG and IgA. This is now a specific reagent, at least in performance of IEP.

(i) Weigh out 2 g of solid CNBr in a fume cupboard. **Note** CNBr is toxic and the next steps (ii and iii) must also be performed in the fume cupboard.

(ii) Suspend 20 g of Sepharose 4B in 50 ml of distilled water and add the CNBr with stirring. Adjust the pH to 11 with 2 M NaOH and maintain over the next 10 min.

(iii) Filter the activated beads on a sintered glass funnel and wash with cold distilled water and then with 0.1 M $NaHCO_3$.

(iv) Wash the beads into a 200 ml glass beaker with the same buffer and allow to settle. Remove the supernatant.

(v) Prepare 250–300 mg of adsorbing protein in 60 ml of 0.1 M $NaHCO_3$ by dialysis and add to the beads. Stir at 4°C for 12–16 h in the covered beaker or rotary mixer, then wash copiously with distilled water, PBS, 0.5 M acetic acid and finally with PBS again until the optical density falls to the PBS blank reading. Adsorbent in PBS with 0.1% (w/v) sodium azide can be stored at 4°C.

(vi) Pack a suitable volume of adsorbent in a column, with PBS as the buffer, using approximately 2 g of the original Sepharose/10 ml serum to be absorbed. Pass the serum down the column at a flow rate of 6 ml/h maximum. Monitor the eluted absorbed protein peak by optical density.

(vii) Between each absorption clear the column of bound antibody by a washing cycle as in Section 9.1.1(vii).

### 9.1.3 *Cellulose carbonate antigen columns*

Microcrystalline cellulose trans, 2,3-carbonate, is available commercially (Sigma). It has much to recommend it as an adsorbent matrix. It has the capacity to couple approximately 40 mg protein (example IgG)/g cellulose under mild, non-denaturing conditions, producing a stable immunoadsorbent with potent binding activity—the general method for preparation (46) is as follows.

(i) Prepare 100 mg of adsorbent antigen in 20 ml 0.1 M phosphate buffer, pH 6.8. To this add, with stirring, 1.6 g of cellulose carbonate. Stir the mixture (rotate) for 24 h at room temperature.
(ii) Wash the adsorbent by centrifugation four times with 10 ml PBS pH 7.4, suspend in 15 ml of PBS and pack in a column (9 × 1 cm) above a 2 ml layer of Sephadex G25. Using IgG as adsorbent 60–70% of the added protein is bound.
(iii) Wash the column with PBS at 10 ml/h until a constant baseline absorbance at 280 nm is reached. Up to 30 ml of serum can be passed through the column at each absorption.
(iv) The column can be recycled for further absorptions by a washing procedure similar to the previous examples.

### 9.1.4 *Red cells as adsorbent carriers*

Antigen-coupled or -coated red cells—that is with their surface coated with a cross-reacting antigen or purified contaminant protein of the immunogen—form a very useful simple method for removing antibodies from sera. For absorption the sera should be first heat-inactivated at 56°C for 30 min to destroy the complement-dependent lytic properties.

The coupling process may be induced through the use of agents such as chromic chloride, glutaraldehyde or tannic acid—these are all useful for the attachment of proteins.

(i) *Chromic chloride method (47,48)*

(1) Wash SRBC in sterile saline three times and centrifuge to a packed volume.
(2) Dilute from a stock solution of chromic chloride to make 1 ml of 0.1% in 0.15 M saline. Add to this 1 ml of protein solution in 0.15 M saline. Usually 1 mg/ml protein is adequate but higher concentrations can be tested.
(3) Add 2 ml of packed red cells to the mixture, cap and rotate at room temperature overnight.
(4) Wash the cells in saline. They can then be used for antibody absorption. The amount of protein-coated cells can be adjusted according to absorption needs.

(ii) *Glutaraldehyde method (49)*

(1) Wash SRBC in sterile saline and centrifuge to a packed volume.
(2) Add 10–20 mg of protein in saline to each 0.4 ml of packed cells (equal volumes), cap and suspend.
(3) Add to the cell/protein mixture 2.5% glutaraldehyde solution. Usually 0.5 ml

will suffice but more may be required according to protein up to 2.0 ml.

(4) Rotate for 1 h at room temperature, then wash the cells in saline copiously before use.

(iii) *The coating of red cells by the tannic acid method of Boyden.* This has been described in Section 6.3.2 above).

(iv) *Antibody-coated cells as immunoadsorbents.* Red cells coated with anti-red cell antibodies of known purified isotype and stabilized with glutaraldehyde can offer a simple means of adsorbing antiglobulin reagents to achieve class-specificity. An example of IgM antibody coating of red cells is given below.

(1) Wash SRBC in PBS and add as a 1% (v/v) suspension to 5 ml of rabbit anti-SRBC IgM antibody (see Section 3.2.3).
(2) Rotate the cells in antibody at 37°C for 45 min and then wash them six times in PBS, with the clumped cells carefully dispersed.
(3) Suspend the packed washed cells at 1% (v/v) in PBS and incubate with 4% glutaraldehyde for 1 h at room temperature with rotation and washed thoroughly before use.

(v) *Absorption of sera with red cells.* Coated red cells as a 1% suspension are mixed with an equal volume of decomplemented sera and rotated in a capped bottle at room temperature for 1 h. The absorbed serum is then recovered by centrifugation. It is convenient to test for any remaining antibody activity by haemagglutination with red cells prepared in the same manner. Slight adjustments to the coating concentration and timing of coupling reactions may be needed to prepare cells for agglutination tests. Absorptions can be repeated with further batches of coated red cells as necessary.

### 9.1.5 *Absorption of sera with bacterial suspensions*

Bacteria harvested from culture in log phase at about $10^8$ organisms/ml are killed by an appropriate method, washed free of culture medium and centrifuged to a button. Heat inactivated antisera can be repeatedly absorbed using each time approximately $10^8$ organisms suspended in 1 ml with rotation for 1 h at room temperature. As an example triple absorption of rabbit anti-*C.jejuni* sera with each of eight other Gram-negative enteric organisms rendered the sera specific to *Campylobacter* spp. with a retained ELISA titre of greater than 1:12 800.

### 9.1.6 *Absorption of antiglobulin sera*

The specificity of antiglobulins is critical in many immunoassays and as immunoglobulin antigens are cross-reactive between isotypes and species the antisera require scrupulous absorption to render them class-, and subclass- and species-specific. One of the commonest mistakes made by the novice is to assume that an antiserum stated to be specific to, for example, rabbit IgG (i.e. does not bind to rabbit IgM or IgA) will not bind to other species IgG—often it will unless absorbed against the IgG of these species.

(i) *Anti-human IgG.* The antiserum is best raised by immunizing with Fc fragment of IgG purified from serum (Section 10). Due to some Fab contamination the sera will usually have anti-light chain antibodies that bind to all immunoglobulin classes and there will also be antibodies to common heavy chain epitopes. An example of the specificity testing of anti-human IgG by ELISA is shown in *Figure 9*.

(1) Absorb with mixed $\kappa$ and $\lambda$ chains (purified Bence Jones proteins of urine), or $F(ab')_2$ prepared by pepsin digestion of IgG.
(2) Absorb with IgM and/or IgA purified myeloma proteins. If both $\kappa$ and $\lambda$ chains are represented, step (1) may not be necessary.
(3) For species-specificity absorb on polymerized mixed serum or column-prepared IgG adsorbents of relevant species.

(ii) *Anti-human IgA and IgM.* The antisera are usually raised against intact molecules as Fc is difficult to prepare whilst absorption with IgG (for $\gamma$ and light chain antibodies) is simple. The two classes can be prepared from myeloma proteins essentially free of each other as they differ greatly in molecular size. Contamination with IgG is removed by use of a Protein A-Sepharose column if necessary (Section 10.4.4). Absorption for class specificity uses IgG plus IgA or IgM and additional light chains. A useful ploy is to polymerize placental cord blood serum that is rich in IgG but lacks IgA and IgM. Absorption for species-specificity uses a euglobulin fraction (by salt precipitation) of sera of relevant species. An example of absorption of anti-human IgM serum and its specificity testing by IEP is shown in *Figure 10*.

(iii) *Anti-human IgG subclasses.* The antisera are raised to Fc or intact molecule of IgG subclass myeloma proteins. Absorption follows steps (1) and/or (2) of (i) and then passage through subclass columns. Many animals respond without adequate subclass definition, and the sera of those that do require very careful absorption. Good precipitating antisera that recognize all of a panel of myeloma proteins confirmed for subclass, without subclass cross-reaction, are very difficult to prepare. The choice of immunized species is important. Fowl species, being phylogenetically distant, have been found to be useful.

The most successful approach to human IgG subclass-specific reagents is through Mabs (24,50).

(iv) *Anti-mouse and -rat Igs.* For a general anti-IgG reagent animals can be immunized with the IgG fraction eluted from a Protein A-Sepharose column (Section 10.6.1), and the serum absorbed against other classes obained as Mabs, or against an IgG-depleted serum.

Anti-IgM serum can be raised using IgM antibodies to the red cells of the immunized species as a red cell/IgM complex, absorbing the serum with an IgG antibody-coated red cell suspension [see Section 9.1.4 (iv)].

For specificity testing a panel of Mabs of all isotypes should be used. These can be applied in GDD tests, or more stringently by PHA, ELISA or IRMA.

Antisera to IgA and to the IgG subclasses are now much more simple to raise since the advent of Mab production in these species which provides a ready supply of affinity-purified molecules which can be used as both immunogens and adsorbing antigens. The easy definition of monoclonal isotypes goes hand-in-hand with the successful preparation of precipitating specific antisera. The sera are raised to intact molecules or their Fc fragments and are column-absorbed with a mixture of other isotypes with both $\kappa$ and $\lambda$ chains. A pure separation of mouse $IgG_{2a}$ for immunization and adsorption can be achieved by physico-chemical means (Section 10.6.1).

Immunization with homogeneous myeloma proteins and Mabs as intact molecules may result in anti-idiotype as well as anti-isotype antibody responses. This is avoided by using Fc fragment. Alternatively the serum can be absorbed with Fab of the immunogen. If anti-idiotype is irrelevant to the antiserum application, isotype specificity should be tested with a different clonal product antigen that has a different variable (Fab) region.

(v) *Anti-Ig polyspecific sera.* The response to intact Ig molecules includes anti-light chain antibodies that bind to all Ig classes and subclasses and the serum is thus an anti-Ig reagent. However, the balance of antibodies will favour the immunizing Ig class unless a mixture of purified intact molecules is used. The better alternative is to mix appropriate proportions of absorbed sera raised to each class and subclass plus anti-$\kappa$ and -$\lambda$ light chain reagents.

## 10. PURIFICATION OF IMMUNOGLOBULINS AND IgG SUBUNITS

There are many requirements for purified immunoglobulins. These are listed in *Table 6.*

### 10.1 **Purification of human IgG**

#### 10.1.1 *Simple method*

(i) Prepare DEAE (Whatman DE52) anion-exchanger as a slurry in 0.01 M phosphate buffer, pH 7.2.
(ii) Dialyse human serum against a 10 vol excess of the same buffer for 24 h at 4°C with five changes of buffer.
(iii) Allow the DEAE slurry to settle in a beaker and take off excess buffer. Using 4–5 ml/ml dialysed serum, transfer the slurry to small beaker or universal and add the serum. Mix thoroughly and stand at room temperature for 1 h with repeated mixing.
(iv) Transfer to a thick-walled centrifuge tube, spin at 4000 r.p.m. for 20 min at 4°C and harvest supernatant as the IgG fraction. Alternatively transfer the slurry to a syringe plugged with filter paper and express the IgG solution by pressure on the barrel. By this procedure serum proteins other than IgG are retained on the DEAE. Some protocols recommend using 0.02 M phosphate at pH 8.0.

#### 10.1.2 *Multi-step method with improved purity*

(i) To whole serum on ice add dropwise with stirring sufficient saturated ammonium sulphate solution (at 0°C) to achieve a 33% saturated (v/v) final concentration.

**Table 6.** Major requirements for purified immunoglobulins and subunits.

| |
|---|
| As immunogens – IgG Fc subunit; IgM, IgA, IgE, IgD as intact purified molecules; $\kappa$ and $\lambda$ chains. |
| As immunoadsorbents – to remove isotype cross-reacting antibodies; for affinity purification of antiglobulins. |
| As standard antigens – for testing antiglobulins and for quantifying immunoglobulins. |
| As the fraction of serum used for antibody labelling. |
| As the starting material for affinity purification of antibodies. |
| As the first step in standardizing antibody reagents for protein and antibody. |
| As an antibody reagent with reduced background properties in immunoassays. |

Stand for 1 h on ice with mixing (avoid frothing) centrifuge at 3000 r.p.m. for 15 min and then wash the precipitate in ice-cold 33% saturated (v/v in distilled water) ammonium sulphate, by centrifugation and resuspension.

(ii) Spin again and remove the supernatant. Dissolve the precipitate (without frothing) in about one-third of the original serum volume of 0.01 M phosphate buffer, pH 7.2 and dialyse the euglobulin solution against this buffer as for serum in Section 10.1.1.

(iii) Prepare a column of DEAE equilibrated with the same phosphate buffer, allowing 15−20 ml as settled volume/5 ml euglobulin solution. Wash the column well and when packing is complete remove excess buffer from the top and layer on the euglobulin. When this has entered the matrix, apply more buffer and attach buffer reservoir.

(iv) Run the column at about 20 ml/h, collecting 5 ml samples. Determine elution of the purified IgG peak by UV monitor or spectrophotometer at 280 nm. Concentration can be determined from the formula: 1 mg/ml. 1 cm = 1.45.

(v) Determine purity of fraction by IEP, using ~ 10 mg/ml protein in the well and antisera to human serum proteins and to IgG (Fc$\gamma$). Use normal human serum (starting material) and purified IgG as standards.

(vi) The column can be washed through with PBS to remove bound proteins and equilibrated with 0.01 M $PO_4$ buffer for re-use.

### 10.2 **Purification of sheep, bovine, goat and rabbit IgG**

The method is in principle the same as for human IgG with variation in the molarity and pH of the phosphate buffer used: sheep and goat, 0.03 M, pH 7.2; rabbit, 0.02 M, pH 7.2 and bovine 0.075 M, pH 6.4.

For preparing the IgG fraction from large volumes of antisera as a routine, the single step purification on DEAE with dialysed serum is recommended. The method achieves sufficient purity for most needs except immunization. Scaling up the volumes of sera and anion-exchanger does not affect yield or purity substantially. Anti-whole serum reagents and anti-IgG reference reagents are required to these species to monitor the success of the procedure.

### 10.3 **Purification of human IgM**

It is not possible to purify IgM completely by a single chromatography procedure as its charge and size characteristics coincide with some other serum proteins. Purification is most easily achieved using as starting material the serum of patients with Waldenström's macroglobulinaemia in which the IgM concentration is greatly increased. Alternatively a euglobulin fraction of normal serum can be prepared as a precipitate in low ionic strength acid buffer as the starting material. This is contaminated with plasminogen and complement components. Fractionation then proceeds by gel-filtration chromatography. Johnstone and Thorpe (11) recommend the use of 6% (w/v) PEG (6000 mol. wt) with Tris-buffered saline, pH 8.0 to prepare the euglobulin fraction of macroglobulin sera in cases where low ionic strength buffer does not yield a precipitate.

### 10.3.1 *Euglobulin fraction*

(i) Dialyse the serum against several changes of 10 vol excess of 2 mM phosphate buffer, pH 6.0 at 4°C.
(ii) Wash the precipitate twice in this buffer and then re-dissolve to the original volume of macroglobulin serum (or about one-tenth of the starting normal serum volume) in 0.01 M Tris–HCl buffer in 0.15 M saline, pH 7.3 at room temperature and centrifuge at 6000 r.p.m. to remove any undissolved material.

### 10.3.2 *Gel-filtration*

(i) Prepare a Sephacryl S300 (superfine, Pharmacia) column, of about 2.5 cm diameter, at room temperature in Tris–HCl–saline buffer, using about 100 ml of gel/2 ml serum or euglobulin, and pack at a flow rate of 200 ml/h with a hydrostatic head of at least 200 cm. A column with valve system, to sustain pressure and to allow insertion of the sample via syringe, is required.
(ii) When packed, with exclusion of air, mix the sample with an equal volume of Blue Dextran, apply it to the column and run the separation at slightly lower pressure.
(iii) Collect eluates in 5 ml samples, eluting the IgM with the coloured marker dye. Sepharose 4B or 6B or Sephacryl S200 can be substituted for Sephacryl S300.
(iv) Concentrate the IgM preparation to about 1 mg/ml by vacuum dialysis against PBS and store at 4°C to avoid precipitation. A yield of about 0.5 mg IgM can be expected per ml normal serum or about ten times this yield from macroglobulin sera.

## 10.4 **Purification of human IgA, IgD and IgE**

The starting material for purification is myeloma sera. It is advantageous to select samples in which the respective isotype of myeloma protein has a slow (cathodic) electrophoretic migration on IEP, as the final purification is achieved by adsorption and elution from anion-exchangers in a buffer gradient. The slower mobility immunoglobulins are eluted with less contamination by molecules migrating in the $\beta$ region. For separation of IgD, the procedures should be performed in the presence of 1 mg/ml of $\epsilon$-amino caproic acid to prevent enzymic degradation of this isotype which is very susceptible to enzymic fragmentation.

(i) Prepare the euglobulin fraction of the myeloma serum by ammonium sulphate precipitation as in Section 10.1.2 (i).
(ii) Dialyse the euglobulin against 0.01 M phosphate buffer, pH 7.2 and apply it to a DEAE column equilibrated with this buffer.
(iii) Elute the isotype on a gradient of phosphate buffer, pH 7.2 from 0.01 M to 0.1 M.
(iv) Remove contaminating IgG molecules by passing the gradient-eluted isotype fraction down a Protein A Sepharose CL-4B (Pharmacia) column. This binds all human IgG molecules except the $IgG_3$ subclass. It is unlikely that this subclass will significantly contaminate the prepared immunoglobulin as it represents only a minor proportion of IgG molecules in normal sera and is even less represented in myeloma sera of IgA, IgD or IgE. When required, $IgG_3$ can be removed by

adsorption to a Sepharose-anti-IgG column, as below:

(v) Use 10 ml of settled volume of CNBr-activated Sepharose to couple the IgG fraction of sheep or rabbit anti-human IgG (Fc$\gamma$-specific), using about 100 mg in 10 ml, following the method described in Section 9.1.2. Wash the immunoadsorbent in PBS, pH 7.2 and prepare as a column in this buffer.

## 10.5 **Purification of human IgG subclasses**

This is achieved by buffer gradient elution of myeloma protein from a DEAE column as described for IgA in Section 10.4 above, using the dialysed euglobulin fraction as the starting material.

Contamination of the eluted subclass with other immunoglobulin classes, or IgG subclasses, is best approached by adsorption to a Protein A-Sepharose CL-4B column which selectively binds $IgG_1$, $IgG_2$ and $IgG_4$. IgG subclasses can be sequentially desorbed using a reducing pH gradient starting with 0.1 M sodium phosphate, pH 8.0 and then increasing proportions of 0.1 M citric acid. $IgG_3$ does not bind to Protein A and this property can be used for further purification.

## 10.6 **Purification of mouse IgG subclasses**

### 10.6.1 *Protein A with normal serum*

Before the advent of Mabs, mouse IgG subclasses had proved notoriously difficult to separate to any degree of purity—a method with normal serum is as follows (11).

(i) Equilibrate a Protein A-Sepharose CL-4B column with sodium phosphate buffer, 0.1 M, pH 8.0. Bring mouse serum to pH 8.0 with 2 M Tris base; add an equal volume of phosphate buffer and apply the mixture to the column. Elute slowly at 5 ml/h with phosphate buffer. The non-bound eluate contains $IgG_3$ which is grossly contaminated with serum proteins including IgM and IgA. An anti-IgG immunoadsorbent is required to purify the $IgG_3$ further, but the yield on acid elution is poor.

(ii) Elute $IgG_1$ from the Protein A column using sodium citrate buffer, pH 6.0.

(iii) Elute $IgG_{2a}$ by reducing the pH further with citrate buffer, pH 4.5.

(iv) Finally, elute $IgG_{2b}$ at pH 3.5 with citrate. Elute the acid fractions into strong neutral buffer and dialyse against PBS, pH 7.2.

(v) An alternative for purification of serum $IgG_{2a}$ is to elute the dialysed euglobulin fraction from DEAE using a phosphate buffer molarity gradient as for human IgA (Section 10.4.2).

### 10.6.2 *Monoclonal antibodies as a source of IgG subclasses*

Purification of mouse IgG subclasses is greatly assisted by the use of Mabs present in ascitic fluid or culture medium. The greatly increased concentration of the isotype in ascitic fluids combined with the homogeneity of the protein in respect of charge (PI), which applies also to culture medium sources, allows greatly improved purity using the Protein A method.

A mixed subclass IgG preparation (lacking $IgG_3$) can be prepared by eluting from a Protein A column at low pH as a single step.

Where Mabs are available in quantity to purified proteins an alternative approach to purifying IgG subclasses (as well as IgM, IgA and IgE) is to use affinity chromatography with antigen immunoadsorbent columns. Consideration of affinity binding and elution is referred to in Chapter 5.

(i) Couple antigen to Sepharose 4B as described previously (Section 9.1.2) and prepare a column in PBS, pH 7.2.
(ii) Prepare an ammonium sulphate euglobulin fraction of ascitic fluid and dialyse against PBS—or dialyse culture medium directly.
(iii) Apply antibody to the column and elute unbound proteins in PBS at 20 ml/h, until absorbance at 280 nm reaches baseline.
(iv) Elute the antibody in alkaline potassium thiocyanate—0.5 M $NH_4OH$ containing 3 M KCNS is recommended, using 2 ml/ml of packed column volume. Chase this with PBS. Dialyse the eluted antibody immediately against PBS.

### 10.7 Preparation of human IgG Fc fragment as immunogen

(i) Dialyse a solution of purified IgG (see Section 10.1) at 10 mg/ml against 0.1 M phosphate buffer, pH 7.0.
(ii) Add cysteine HCl (1 mg/10 mg IgG) and EDTA (0.5 mg/10 mg IgG), followed by papain (1 mg/100 mg IgG). Incubate the mixture for 4 h at 37°C.
(iii) Perform IEP (Chapter 6) on the product, using an antiserum to IgG whole molecule to determine the extent of digestion and the presence of Fc fragment which runs anodal to intact molecule and forms a crossed arc with this.
(iv) When digestion is substantial ($IgG_3 > IgG_1 < IgG_4 < IgG_2$), freeze the digest to inhibit further digestion and then dialyse it at 4°C against 0.01 M phosphate buffer, pH 8.0.
(v) Equilibrate DEAE (DE-52) in the 0.01 M phosphate buffer and prepare a column. Apply the digest to the column and run a gradient in phosphate buffer from 0.01–0.3 M at pH 8.0.
(vi) The first peak is Fab (useful as adsorbent); the second peak is predominantly Fc with some undigested IgG (mainly $IgG_2$).
(vii) Remove contaminating Fab by passing the second peak through a Protein-A Sepharose CL-4B column which binds the Fc and intact molecules of all but $IgG_3$ but allows the Fab to pass through. Elute the Fc and IgG with 3 M KCNS and dialyse the collected fraction against PBS.
(viii) Finally, separate IgG and Fc on a Sephadex G200 column in PBS—the IgG comes through in the first peak and pure Fc in the second.

For immunization, as the Fc migrates anodally it will run into the agarose of a 2D-IEP plate (Chapter 6) containing antibodies to IgG, to form a discrete precipitation peak. The Fc peak of an unseparated IgG digest can be excised and used as the source of purified Fc immunogen.

## 11. QUALITY CONTROL OF ANTIBODIES

Descriptions of standardization of immunoassays usually consider only the importance of standard antigen for giving reliable data with minimum discrepancy between laboratories. However, the quality of antiserum plays an equally important role and must be attended to as a first priority. This applies whether antisera are home-produced or purchased from a commercial source. No manufacturer or retailer is able to test their catalogue of antibody specificities under all possible laboratory applied conditions or for all possible applications and it rests with the user to test adequately the intrinsic properties of the reagent.

### 11.1 Guidelines

(i) Use a standard antigen for testing sera. According to the intended use this may be a purified molecule in solution, mixed antigen molecules, microorganisms, cells, tissues etc.

(ii) Use a reference antiserum for comparison wherever possible (see *Figures 8* and *10* as examples).

(iii) Always incorporate tests appropriate to the intended use of the reagent.

(iv) Test each bleed of each animal separately and only pool sera that are entirely adequate.

### 11.2 Sequence of control measures

(i) Test for specificity, determine absorptions required, perform these and retest (see *Table 5*).

(ii) Measure the titre of an antibody reagent; a prepared concentrated IgG fraction can be added to serum to increase the ratio of antibody to total protein. Alternatively the whole serum can be prepared as an IgG fraction at required antibody concentration. This has the advantage of increasing still further the antibody to protein ratio, reducing the amount of protein added to assay systems, and reducing background readings. It also allows exact adjustments of antibody concentration between successive batches of the same antibody specificity. It is advisable to keep some antibodies of above required titre in reserve to 'improve' later reagents. An antibody reagent for use in precipitation tests should contain at least 1–2 mg specific antibody/ml: 5–10 mg/ml is preferable to reduce the amount of protein added. The purity of the IgG fraction of prepared antibodies should be tested. When this is satisfactory, the total protein can be estimated by optical density and the antibody:total IgG ratio determined as a bench mark.

(iii) Avidity: the requirement for high avidity varies greatly between assay systems—being critical in fluid phase RIAs and where low backgrounds are required, including cell and tissue staining and Western immunoblots. The hall-mark of a high avidity antibody reagent is that it retains optimal binding to high dilution and then falls rapidly to negative values through inadequate supply as opposed to dissociation in antigen excess. One measure of this property that can be easily determined is the slope angle of the binding curve in an antibody dilution series against constant antigen. Sustained optimal binding followed by a precipitous linear decline is evidence of high avidity. The slope can be defined as percentage binding decline (e.g. optical density in ELISA, precipitated radioactivity etc.) over a single dilution step in the linear decline phase. Sharply defined precipitation lines and rings in gel precipitation assays is also a good indication of high avidity of a polyclonal antiserum.

Assays which measure formation of immune complexes in solution, such as nephelometry, also measure avidity of antisera. High avidity sera form stable complexes rapidly; such reagents are critically important in these assays as sensitivity depends upon this property. Only a small proportion of sera raised in animals have ideal nephelometric properties.

(iv) Affinity: this can only be determined for antibodies to small haptenic groups using equilibrium dialysis (see ref. 51).

(v) Isotype: although in rabbits and sheep the antibodies prepared by FCA immunization are of homogeneous isotype (IgG and $IgG_1$), in other species, such as rats, mice and guinea pigs, the IgG subclass pattern may be restricted. When assays demand use of labelled antiglobulin it will be necessary to define the responding isotypes to ensure that an appropriate antiglobulin specificity is applied.

## 12. ACKNOWLEDGEMENTS

We thank R.Drew for advice on immunoglobulin purification and strategies of immunization for antiglobulins and their absorption; Dr A.R.Bradwell for *Figure 7*. We are indebted to Dr L.Piddock of the Department of Clinical Microbiology for assistance in preparing *E.coli* antigens and their SDS−PAGE separation. Finally we wish to thank Mrs F.O'Reilly for typing the manuscript, and the staff of Educational Services for the figures.

## 13. REFERENCES

1. Borek,F. (1977) In Sela,M. (ed.), *The Antigens*, Vol. IV. Academic Press, New York, p. 369.
2. Warren,H.S., Vogel,F.R. and Chedid,L.A. (1986) *Annu. Rev. Immunol.*, **4**, 369.
3. Adam,A. (1985) *Synthetic Adjuvants*. John Wiley, New York.
4. Freund,J. (1947) *Annu. Rev. Microbiol.*, **1**, 291.
5. Glenny,A.T., Pope,C.G., Waddington,H. and Wallace,U. (1926) *J. Pathol. Bacteriol.*, **21**, 31.
6. Munoz,J.J., Arai,H., Bergman,R.K. and Sadowski,P. (1981) *Infect. Immunol.*, **33**, 820.
7. Allison,A.C. and Gregoriadis,G. (1974) *Nature*, **252**, 252.
8. New,R.R.C., Theakston,R.D.G., Zumbuehl,O., Iddon,D. and Friend,J. (1984) *New Engl. J. Med.*, **311**, 56.
9. New,R.R.C., Theakston,R.D.G., Zumbuehl,O., Iddon,D. and Friend,J. (1985) *Toxicon*, **23**, 215.
10. Kotani,S., Watanabe,Y., Kinoshita,F., Shimono,T., Morisaki,T., Shiba,T., Kusumoto,S., Tarumi,Y. and Ikenaka,K. (1975) *Biken J.*, **18**, 105.
11. Johnstone,A. and Thorpe,R. (1982) *Immunochemistry in Practice*. Blackwell Scientific Publications, Oxford.
12. Parker,C.W. (1976) *Radioimmunoassay of Biologically Active Compounds*. Prentice Hall, New Jersey.
13. Williams,C.A. and Chase,M.W. (eds) (1967) In *Methods in Immunology and Immunochemistry*, Vol. 1. Academic Press, New York, p. 197.
14. Mäkelä,O. and Seppälä (1986) In Weir,D.M. (ed.), *Handbook of Experimental Immunology*, Vol. 1. Blackwell Scientific Publications, Oxford, 4th edition, p. 3.1.
15. Gray,D., Chassoux,D., Maclennan,I.C.M. and Bazin,H. (1985) *Clin. Exp. Immunol.*, **60**, 78.
16. Wills,P.J. (1979) Ph.D. Thesis. University of Birmingham, UK.
17. Erlanger,B.F., Borek,F., Beiser,S.M. and Lieberman,S. (1957) *J. Biol. Chem.*, **228**, 713.
18. Murphy,G.M., Edkins,S.M., Williams,J.W. and Catty,D. (1974) *Clin. Chim. Acta*, **54**, 81.
19. Catty,D., Raykundalia,C. and Houba,V. (1983) *WHO Bench Manual*, Part II, IMM/PIR/83:1.
20. Inman,J.K. (1975) *J. Immunol.*, **114**, 704.
21. Weir,D.M. (ed.) (1986) *Handbook of Experimental Immunology*, Vol. 1. Blackwell Scientific Publications, Oxford, 4th edition.

22. Sternick,J.L. and Sturmer,A.M. (1984) *Hybridoma,* **3**, 74.
23. Knudsen,K.A. (1985) *Anal. Biochem.,* **147**, 285.
24. Lowe,J., Bird,P., Hardie,D., Jefferis,R. and Ling,N.R. (1982) *Immunology,* **47**, 329.
25. Ling,N.R., Elliot,D. and Lowe,J. (1987) *J. Immunol. Methods*, in press.
26. Hames,B.D. and Rickwood,D. (eds) (1981) *Gel Electrophoresis of Proteins—A Practical Approach.* IRL Press, Oxford.
27. Filip,C., Fletcher,G., Wulff,J.L. and Earhart,C.F. (1973) *J. Bacteriol.,* **115**, 717.
28. Dresser,D.W. (1986) In Weir,D.M. (ed.), *Handbook of Experimental Immunology.* Vol. 1. Blackwell Scientific Publications, Oxford, 4th edition, p. 8.1.
29. Williams,C.A. and Chase,M.W. (eds) (1967) *Methods in Immunology and Immunochemistry*, Vol. 1. Academic Press, New York, pp. 197–306.
30. Theakston,R.D.G., Zumbuehl,O. and New,R.R.C. (1985) *Toxicon,* **23**, 921.
31. Herbert,W.J. (1967) In Weir,D.M. (ed.), *Handbook of Experimental Immunology*, Appendix 4. Blackwell Scientific Publications, Oxford, p. A 4.1.
32. Poxton,I.R. and Blackwell,C.C. (1986) In Weir,D.M. (ed.), *Handbook of Experimental Immunology.* Blackwell Scientific Publications, Oxford, 4th edition, p. 4.1.
33. Daniel,T.M. and Janicki,B.W. (1978) *Microbiol. Rev.,* **42**, 84.
34. Seibert,F.B. (1949) *Am. Rev. Tuberc. Pulm. Dis.,* **59**, 86.
35. Boyden,S.V. (1951) *J. Exp. Med.,* **93**, 107.
36. Pearson,T.W. and Clarke,M.W. (1986) In Weir,D.M. (ed.), *Handbook of Experimental Immunology.* Blackwell Scientific Publications, Oxford. 4th edition, p. 6.1.
37. Penner,J.L. and Hennessy,J.N. (1980) *J. Clin. Microbiol.,* **12**, 732.
38. Catty,D., Humphrey,J.H. and Gell,P.G.H. (1969) *Immunology,* **16**, 409.
39. Parson,M., Herzenberg,L.A., Stall,A.M. and Herzenberg,L.A. (1986) In Weir,D.M. (ed.), *Handbook of Experimental Immunology.* Vol. 3. Blackwell Scientific Publications, Oxford. 4th edition, p. 97.1.
40. Kelus,A.S. and Gell,P.G.H. (1967) *Progr. Allergy,* **11**, 141.
41. Roth,C., Somme,G., Goufeon,M.L. and Theze,J. (1985) *Scand. J. Immunol.,* **21**, 361.
42. Ling,N.R., Elliott,D. and Lowe,J. (1987) *J. Immunol. Methods*, in press.
43. Hole,N.J., Catty,J.P. and Catty,D. (1986) *Mol. Immunol.,* **24**, 75.
44. Praputpittaya,K. (1986) Ph.D. Thesis, University of Birmingham, UK.
45. Eichmann,K. and Kindt,T.J. (1971) *J. Exp. Med.,* **134**, 532.
46. Kennedy,J.F., Catty,D. and Keep,P.A. (1980) *Int. J. Biol. Macromol.,* **2**, 137.
47. Ling,N.R., Stephens,G., Bratt,P. and Dhaliwal,H.S. (1979) *Mol. Immunol.,* **16**, 637.
48. Gold,E.R. and Fudenberg,H.H. (1967) *J. Immunol.,* **99**, 859.
49. Avrameas,S., Taudou,B. and Chuilon, (1969) *Immunochemistry,* **6**, 67.
50. Jefferis,R. *et al.* (1985) *Immunol. Lett.,* **10**, 223.
51. Steard,M. (1986) In Weir,D.M. (ed.), *Handbook of Experimental Immunology.* Vol. 1. Blackwell Scientific Publications, Oxford, 4th edition. p. 25.1
52. Mage,R.G. (1981) *Contemp. Top. Mol. Immunol.,* **8**, 89.
53. Herzenberg,L.A. and Herzenberg,L.A. (1978) In Weir,D.M. (ed.), *Handbook of Experimental Immunology.* Blackwell Scientific Publications, Oxford, 3rd edition. p. 39.

CHAPTER 3

# Murine monoclonal antibodies

GEOFFREY BROWN and NOEL R.LING

## 1. BASIC PRINCIPLES FOR THE GENERATION OF ANTIBODY-PRODUCING CELL LINES

In the 1950s, Burnet, Jerne, Lederberg and Talmage postulated that each B-lymphocyte is pre-determined to make only one particular antibody molecule which is expressed at the cell surface as a receptor for antigen. This is now well proven and when an animal is immunized with antigen, this leads to clonal expansion of those B-lymphocytes which recognize antigen and their differentiation to antibody-secreting cells. If individual antigen-responding cells from lymphoid tissue of an immunized animal could be propagated continuously *in vitro* the culture supernatant would contain homogeneous antibody molecules (monoclonal antibody) which recognize only one or a few closely related antigens. Unfortunately the progeny of an individual responding lymphocyte cannot be grown long-term in culture to produce large amounts of monoclonal antibody (Mab). This was achieved in 1975 by Kohler and Milstein (1), who first immortalized antibody-secreting lymphocytes by fusing them with cells from a continuously growing cell line and then cloned individual hybrid cells to produce lines of cells (hybridomas) each of which secrete one particular antibody molecule.

It has been known for a long time that cells of different types and different species may be fused to form hybrid cells. However, highly differentiated characteristics are not retained unless the two partners are cells of similar lineage at approximately the same stage of differentiation (2). An antibody-secreting cell line is produced by fusing together an antibody-secreting cell from lymphoid tissue of an immunized animal and a cell from a plasmacytoma cell line which represents a similar differentiation stage. The resultant hybridoma retains the ability of one parent to secrete a particular antibody molecule and the continuous growth characteristics of the other parent (the plasmacytoma). The technique has been highly successful with mouse and rat cells; there has been less success with human hybridomas and little work has been done on other species. An alternative approach for the generation of human antibody-producing cell lines is to use Epstein–Barr virus (EBV) to transform the B cell population of an immune individual (3). Cells producing high levels of antibody are then selected and cloned (see Chapter 4). Only the hybridoma technique will be considered here. There are many reviews on this subject (see refs 2,4–6).

## 2. EQUIPMENT AND MATERIALS REQUIRED

Good tissue culture facilities, including a 5% $CO_2$ gassed incubator, are essential and a laminar flow hood is highly desirable. At this point it is worth noting that good basic aseptic technique and some experience in growing cell lines are important attributes. An inverted microscope (e.g. Olympus CK model fitted with a broad flat stage, a ×10 objective and binocular WF15x eyepieces) is needed for inspection of cultures. Apart from liquid nitrogen containers for storage of plasmacytomas and hybridomas, little else is required for production of monoclonal antibodies. Assay systems may need to be refined or developed for screening procedures and are considered later. Inbred strains of mice (usually BALB/c) or rats (Lou) must be available for immunizing and large colonies will be needed if hybridomas are to be grown up as ascitic tumours.

## 2.1 Materials

### 2.1.1 *Plasmacytoma cell lines*

The plasmacytoma lines used have a deficiency in the enzyme hypoxanthine-guanine phosphoribosyl transferase (HGPRT) and have been selected by growing cells in medium containing thioguanine ($2 \times 10^{-5}$ M). This means that HGPRT-deficient cells are unable to synthesize DNA when cultured in medium containing hypoxanthine, aminopterin and thymidine (HAT medium). The lymphocyte parent complements this deficiency during the fusion process and thus only hybridoma cells are able to grow in HAT medium.

Plasmacytoma cell lines are maintained routinely in standard single strength RPMI-1640 medium with L-glutamine (Gibco Bio-cult) supplemented with 10% fetal calf serum (FCS), the antibiotics penicillin (100 U/ml) and streptomycin (100 μg/ml) and thioguanine (20 μM). A 100× stock solution of penicillin and streptomycin at the correct concentrations is available commercially (Gibco Bio-cult). The medium is supplemented with thioguanine to eliminate any revertant cells containing 'normal' levels of HGPRT which, like the hybridoma cells, will not be killed in HAT medium.

(i) Prepare the thioguanine as a 100× stock solution at 2 mM as follows. Dissolve 33.44 mg of anhydrous thioguanine in 100 ml of distilled water. Add 1 M NaOH, as necessary, to dissolve the thioguanine and adjust the pH to 9.5 with acetic acid. Filter using Millipore filters (0.2 μm pore size), aliquot and store at −20°C.

(ii) To prepare a bottle of medium for maintaining the plasmacytoma culture, combine the following stocks:

  500 ml bottle of RPMI-1640 medium containing L-glutamine
  50 ml of heat-inactivated FCS (final concentration ~9% FCS)
  5.5 ml of stock penicillin/streptomycin
  5.5 ml of stock thioguanine

The HAT-sensitive BALB/c plasmacytoma P3-NSI/Ag4-1 (commonly referred to as NSI) is a non-secreting variant of the MOPC21 cell line (4). It grows vigorously *in vitro* in RPMI-1640 medium containing 10% FCS and will also grow in BALB/c mice. It synthesizes but does not secrete light chains of immunoglobulin. The HAT-sensitive BALB/c plasmacytoma X63/Ag8.653, a variant of the X63/Ag8 line, does not synthesize heavy or light chains of immunoglobulin (4). Cells of both lines fuse well but hybridomas produced with NSI grow somewhat more vigorously.

### 2.1.2 *Fusogens*

The fusogen used in all the early work was UV-inactivated Sendai virus. This has now been replaced by polyethylene glycol (PEG) which is prepared as a 50% (w/v) solution. The 50% PEG is made up in water rather than the more usual culture medium because it is hypertonic at this concentration. It is also slightly acid; 1 mM NaOH is sufficient to neutralize it.

(i) Weigh out 8 g of PEG 1500 into a glass universal.

(ii) Prepare 1 mM NaOH in distilled water and add one drop of phenol red to it (red–purple tint). Add 8 ml of this solution to the PEG and place at 37°C to allow the PEG to dissolve.

(iii) Sterilize by autoclaving the loosely capped universal and check that liquid is not lost during this process by marking the level of the solution on the universal. If some liquid is lost, return the volume to the original mark by adding 1 mM sterile NaOH.

(iv) Allow to cool, prepare 3 ml aliquots in sterile vials and store at 4°C.

2.1.3 *HAT medium*

This medium is used to select hybridoma cells and grow them routinely and consists of standard single strength RPMI-1640 medium containing L-glutamine, 20% FCS, penicillin (100 U/ml)/streptomycin (100 μg/ml), 100 μM hypoxanthine, 16 μm thymidine and 0.5 μM methotrexate.

Prepare the following stock solutions.

(i) *Hypoxanthine and thymidine.* This solution is prepared as a 100× combined stock at 10 mM hypoxanthine and 1.6 mM thymidine. Dissolve 408 mg of hypoxanthine in 100 ml of distilled water by stirring and adding 1 M NaOH dropwise until all the hypoxanthine has dissolved. Dissolve 114 mg of thymidine in 100 ml of distilled water. Combine these two solutions and make the volume up to 300 ml with distilled water. Adjust to pH 10.0 with HCl or NaOH, Millipore filter to sterilize, aliquot and store at −20°C.

(ii) *Aminopterin.* Aminopterin (100× stock at 50 μM) is prepared from medical ampoules of methotrexate which contain 2 ml at 25 mg/ml. Take care when removing methotrexate from glass ampoules as this substance is harmful. Dilute 0.9 ml to 1 litre with distilled water and adjust to pH 7.5 with 0.01 M NaOH or 0.01 M HCl. Filter to sterilize, aliquot and store at −20°C.

(iii) *2-Mercaptoethanol.* 2-Mercaptoethanol at a concentration of 50 μM can be added to HAT medium to assist the growth of hybridoma cells and must be added when cloning hybridoma cells. To prepare a stock solution add 0.06 ml of the mercaptoethanol solution to 10 ml of standard RPMI-1640 medium. Filter to sterilize, aliquot and store at −20°C.

To prepare a bottle of complete HAT medium combine the following stocks.

500 ml bottle of RPMI-1640 medium with L-glutamine
100 ml bottle of heat-inactivated FCS (which gives ~16% FCS)
6 ml of stock penicillin/streptomycin
6 ml of stock hypoxanthine/thymidine
6 ml of stock methotrexate
12 drops (0.36 ml) of 2-mercaptoethanol are added using a Pasteur pipette as indicated above.

2.1.4 *Dulbecco's medium*

Dulbecco's medium is made up from a powder (Gibco Bio-cult) and may be used instead of RPMI-1640 medium or a combination of media may be used. 'Alkaline Dulbecco or RPMI-1640' medium used at the fusion stage, in one protocol, is Dulbecco's and RPMI-1640 medium which have been allowed to go alkaline in a universal container by equilibration with the atmosphere. This is most easily achieved by simply loosening the lid for several days before use.

2.1.5 *Serum supplement*

This is usually FCS which is inactivated by heating in a water bath at 56°C for 30 min. Different batches of FCS vary considerably in growth-promoting quality and it is vital to choose a good batch. The best and simplest test is to perform limiting dilutions (see Section 6.4) of a hybridoma in medium containing the samples of sera under test. An additional supplement of horse serum (5%) is often beneficial if the FCS is not top grade. Human cord blood serum has also been recommended as a supplement (7). It is, of course, inadvisable to use a serum which contains an antigen to which Mabs are directed (e.g. human immunoglobulin or a species which cross-reacts, when it is desired to select Mabs to immunoglobulin).

2.1.6 *Hepes-buffered RPMI medium (H-RPMI)*

This is RPMI-1640 medium which has been buffered with 4.6 g/l Hepes (*N*-2-hydroxy-ethylpiperazine-*N'*-2-ethanosulphonic acid) instead of the usual 2 g/l sodium bicarbonate. It is supplemented with 2% FCS (H-RPMI−2% FCS). It is ideal as a wash reagent or suspending reagent during handling of cells outside the incubator because it retains its pH under atmospheric conditions. It is unsuitable as a growth medium.

## 3. IMMUNIZATION OF MICE

Animals, 3−4 months old, of either sex may be used. The immunization schedule used will depend upon the nature of the immunogen just as it does in the preparation of a polyclonal antiserum. It is important to determine that mice used for fusion experiments have responded to the immunogen. A test bleed is taken from the tail vein of mice usually 1 week after concluding the immunization protocol. If an adequate titre of anti-body is present in the serum, a final boost(s) is given 3−4 days before removal of the spleen for cell fusion. It is sometimes difficult to interpret the test bleed result because such a complex mixture of antibodies is present in the serum. It does not follow that all the antibody is being produced by secreting cells in the spleen or that antigen-reactive cells of the same clones will dominate the response when antigen is re-introduced. However, the test bleed does indicate whether or not the right sort of antibodies are present.

### 3.1 **Immunization with cell surface (and other particulate) antigens**

For cell surface antigens the following immunization schedules can be used. The choice may be dictated by the number of cells available for immunization.

(i) Two intraperitoneal (i.p.) injections of $10^7$ whole cells [in phosphate-buffered saline (PBS)] given 4 weeks and 3−4 days pre-fusion may be adequate.

(ii) We prefer to follow the primary dose of $10^7$ whole cells, given i.p. 3–4 weeks pre-fusion, with i.p. boosts 7 days and 3 days pre-fusion. These mice are test bled 1 week prior to the 7 days pre-fusion boost.

(iii) Alternatively, we immunize mice by injecting $10^7$ cells (in PBS) i.p. on days 0, 14 and 21. Seven days later test bleed the mice. Mice with high levels of antibody are injected intravenously (i.v.) with $10^7$ cells 3 days pre-fusion.

(iv) When mice are immunized with human cells a strong response against species antigens is evoked. In order to bias the response of mice used in fusion experiments towards antigens expressed selectively by a particular cell type, an antibody coating strategy can be used. For example, this approach has been used successfully to produce Mabs antibodies which identify antigens selectively expressed by human myeloid cells. Myeloid cells were coated with mouse antibody to human peripheral blood leukocytes prior to immunizing mice for fusion experiments (8). The rationale for this strategy is that by coating the cells in this manner immunogenic surface proteins shared by all human cell types are masked.

Antigen injected i.p. or i.v. will reach the spleen and, since this tissue is normally the source of the antibody-producing cells used for fusion experiments, it is important that it should do so. In a number of studies the pre-fusion boost is given i.v. and this may cause anaphylaxis. It is not necessary to give this injection i.v. but to avoid the possibility of shocking the animal it should be preceded by an i.p. injection given several hours earlier (i.p. in the morning and i.v. in the afternoon).

### 3.2 **Immunization with soluble antigens**

For soluble antigens many different protocols have been used. For example, give a primary injection of 100 μg antigen emulsified in Freund's complete or incomplete adjuvant subcutaneously, in 0.1 ml, followed by i.p. boosts of 100 μg of antigen in PBS (0.5 ml) at monthly intervals. A test bleed is taken to ensure an adequate titre of antibody is present in the serum and final boosts are given, i.p. in the morning and i.v. in the afternoon, 3–4 days pre-fusion.

Some workers prefer to avoid the use of Freund's adjuvant. It may indeed stimulate the development of granulomas, the cells of which may increase the background growth in post-fusion cultures. In preference, alum-precipitated antigen with *Bordetella pertussis* as adjuvant (9) can be used. Another technique which has been used successfully with immunoglobulins is to couple them to plastic particles (10) and immunize i.p. without adjuvant. Relatively small peptides may need to be coupled to a carrier protein but may be immunogenic if administered in Freund's complete adjuvant.

## 4. FUSION METHODOLOGY

There have been many modifications to the original Kohler and Milstein technique (1). Popular procedures are those of Galfré (11) and Kennett (5) but each centre tends to develop its own preferred protocol.

Two fusion techniques will be described before discussion of some of the more important variables. In this department the two approaches work equally well and illustrate that wide variations in the protocols are possible without impairment of success in fusions. In particular, in the first procedure, post-fusion, cells are plated and handled in 96-well microtitre plates. In the second procedure hybridoma colonies are grown in 2 ml cultures in 24-well culture plates. In this respect choice of culture plate might relate as much to convenience of handling as to familiarity with cell handling procedures and availability of equipment.

Success in performing fusion experiments will depend on careful and appropriate handling of cells during the fusion process (see Section 4.3). If the initial attempt is unsuccessful then repeat the protocol you first chose to use. Experience gained at the first attempt should ensure a more careful second experiment. If you still have problems in growing hybrid cells then try the second protocol. This may avoid a problem of which you are unaware (see Section 4.3).

## 4.1 **Protocol A**

It is convenient and advantageous to perform two fusion experiments simultaneously with individual spleens from two mice. Prepare the following materials before fusion.

(i) Sample the plasmacytoma line by resuspending the cells growing in the thioguanine medium. In the case of NSI cell line, the cells are not readily resuspended by inverting the bottle and need to be blown off the surface by sucking up the medium in a Pasteur pipette and ejecting it at the cell layer. Remove four drops, add one drop of 1% trypan blue and enumerate viable cells. For each of the two fusions to be performed spin down $10^7$ viable cells in a round-bottom capped 30 ml Pyrex tube at 500 *g* for 5 min. Pour off the supernatants, resuspend the cells in 20 ml of H-RPMI−2% FCS and leave at 4°C.

(ii) 4 × 10 ml syringes, each containing 10 ml of H-RPMI−2% FCS and fitted with 26-gauge needles. Store with a shield inside the sterile package in the hood.

(iii) 2 × 1 ml syringes, each containing 0.7 ml of 50% PEG in 1 mM NaOH. Store in a 37°C incubator.

(iv) 2 × 10 ml of RPMI-1640 medium in two universal containers. Store at 4°C.

(v) 2 × 20 ml of RPMI-1640 medium in two universal containers. Store at 37°C.

(vi) Alkaline RPMI-1640 medium. A small volume in a universal container. Store in the hood.

(vii) 2 × 30 ml of complete HAT medium in bottles. Store in a 37°C incubator.

(viii) On the bench: stopwatch, 70% ethanol, sterile instruments, absorbent paper, balance tube containing 21 ml of water, 37°C bath, 1 ml syringes, 15-gauge needles, two Petri dishes containing 1 ml of H-RPMI−2% FCS.

(ix) In the hood: two Petri dishes, Pasteur pipettes, pipettes.

(x) 96-Well microtitre culture trays. These may be used straight from stock. However, we prefer to add one drop of a mixture of equal parts of horse serum and HAT medium to each well and to store the trays at 4°C while the fusion is proceeding.

### 4.1.1 *Fusion procedure A (based on two fusions)*

At this stage the mice have been previously test bled and sera analysed and shown to contain acceptable levels of antibody against the immunogen.

(i) Kill the mouse by breaking the neck and shower it with 70% ethanol from a wash-bottle. Remove excess fluid by wiping with absorbent paper. Place the mouse onto a dissection board which has been cleaned with 70% ethanol and slit it open with sterile instruments to reveal the spleen. Use fresh scissors and forceps to remove the spleen, place it in a puddle of medium in a small sterile Petri dish and transfer it to the sterile hood.

(ii) After rinsing, transfer the spleen to a fresh Petri dish and proceed to blow out the lymphoid cells using two 10 ml syringes fitted with 26-gauge needles, each containing 10 ml of H-RPMI−2% FCS. To achieve this use one syringe to anchor the spleen, puncture it with the needle of the other and blow out fluid into successive small areas of spleen using the fluid from first one and then the other syringe.

(iii) Remove the spleen carcass (looking slightly lighter but otherwise unchanged) and transfer the cell suspension with a Pasteur pipette to a 30 ml tube leaving behind any small lumps of tissue which may have settled on the base of the dish. Leave in a rack in the hood. Repeat the procedure with another mouse spleen.

(iv) Place four drops of each cell suspension into a well of a (non-sterile) microtitre tray. Set the labelled spleen tubes and the plasmacytoma tubes (removed from 4°C refrigerator) to centrifuge at 500 *g* for 5 min. Meanwhile add a drop of 1% trypan blue to each cell suspension in the microtitre tray, mix the cells using a Pasteur pipette and transfer a sample to a counting chamber. Count the viable lymphocytes and lymphoblasts. The total cell yield per spleen is usually $0.6-1.2 \times 10^8$ and viability approximately 90%. In most experiments $10^8$ spleen cells are fused with $10^7$ plasmacytoma cells.

(v) When centrifugation is complete remove the tubes from the centrifuge and carefully pour off the supernatants into a waste pot in the hood. Flick the tubes to disperse cell deposits. Add 10 ml of cold RPMI-1640 medium to each spleen tube and pour each spleen cell suspension into one of the tubes containing plasmacytoma cells. Spin at 400 *g* for 7 min and during this time transfer the 37°C bath, which should be clean, to the hood. Aspirate the supernatants completely with a Pasteur pipette and add five drops of alkaline RPMI-1640 medium to the cell pellets, flick the tubes to disperse the cells.

(vi) The cell fusion procedure should take 35 min to perform. Start the stopwatch, set to 35 min and proceed as in *Table 1*. Repeat the same procedure for tube no. 2 (start as at 29.00 for tube no. 1). When centrifugation of tube no. 1 is complete remove to the hood and centrifuge tube no. 2 which should now be ready.

(vii) Pour off the supernatant of tube no. 1 and pour 30 ml of warm HAT medium onto the cell pellet. Replace the cap and invert the tube to resuspend and disperse the pelleted cells. Set out and label six microtitre trays for each fusion. Use a Pasteur pipette to dispense the cells, one drop per well, into six trays each with 96 wells. When all the trays are complete transfer them in a stack to a gassed incubator which is humid. Repeat the procedure with cells from fusion no. 2.

It is particularly important during the early stages of culture when the total volume in the well is only approximately 0.07 ml that an atmosphere saturated with moisture is maintained in the $CO_2$ incubator, otherwise evaporation will occur. On day 5, 0.15 ml of HAT medium is added to each well. This is done with a multichannel pipette fitted

**Table 1.** Timed procedure for cell fusion.

| *Time (min)* | *Sequence of steps* |
|---|---|
| 34.00–29.00 | Coat the cell deposits around the tubes by holding the tubes horizontally and rotating them (both tubes, alternately). |
| 29.00–25.30 | Place tube no. 1 in the 37°C bath and remove occasionally, wipe and rotate horizontally. |
| 25.30 | Remove the PEG syringe from the incubator to the hood. Transfer RPMI-1640 medium (20 ml) to the bath. |
| 24.30–23.30 | Remove tube no. 1 from the bath, wipe and rotate horizontally. |
| 23.00 | Rapidly add the PEG (0.7 ml) to tube no. 1. |
| 23.00–21.00 | Rotate tube no. 1 horizontally. The entire wall of the tube should be coated with cells. |
| 21.00–20.30 | Put tube no. 1 in a rack outside the bath and remove the top. Place tube no. 2 in the 37°C bath. Place a Pasteur pipette fitted with a bulb into the tube containing 20 ml of warm RPMI-1640 medium in the bath. |
| 20.30–19.30 | Rapidly add ~1 ml of warm RPMI-1640 medium to tube no. 1 and rotate horizontally. |
| 19.30–18.30 | Rapidly add another 1 ml of warm RPMI-1640 medium to tube no. 1 and again rotate to mix. |
| 18.30–15.30 | Remove the cap from tube no. 1, place the tube in the bath and slowly add the remainder of the RPMI-1640 medium with mixing in the bath. |
| 15.00–0.00 | Spin the tube at 300 *g* for 15 min (using the 21 ml balance tube as companion). |

with sterile polythene tips, the medium having been poured into a sterile reservoir of suitable size (an autoclaved aluminium pie dish can be used).

When many of the wells are clearly yellow due to acid-producing cell growth, the entire medium is changed to eliminate background antibody produced by residual immunoglobulin-secreting spleen cells. This is done by drawing out the end of a Pasteur pipette to a fine tip and then making a right-angle bend about 3 cm higher up. The drawn pipettes are sterilized by autoclaving in aluminium containers. Place some cotton wool at the bottom of the container to avoid damaging the tips. The Pasteur pipette is connected to a suction pump and the entire fluid contents of a tray are aspirated and replaced by 0.2 ml of HAT medium (sometimes containing 5% horse serum) using the multichannel pipette. All the trays are treated in this way and replaced in the incubator until, usually 2 days later, many of the wells have again turned yellow. It is now time to test the supernatants for antibody. This may be done by testing individual wells as they become yellow or by leaving trays until many wells have turned yellow and then testing all the supernatants. The first procedure, which may be preferred if a labour-intensive assay system is used, consists of removing almost all the supernatant with a Pasteur pipette, transferring it to a numbered test tray and replacing the medium before proceeding to the next yellow well. It is obviously vital to avoid contaminating cells of one well with those of another. A single Pasteur pipette may be sterilized by squirting boiling water back and forth and is cooled by sucking up cold medium before proceeding to the next sample. When a simple assay system such as passive haemagglutination is used it is simpler and more informative to sample the entire tray using the multichannel pipette. Screening strategies are discussed in Section 5.

Trays should be inspected daily under the inverted microscope. For the first 2 days nothing is usually seen other than single cells of the size and appearance of the plasmacytoma against a background of assorted spleen cells. By day 4 small colonies of hybridomas should have begun to appear in some wells if the fusion is to be successful. By day 5 when more medium is added, large colonies should be present in some of the wells and by day 7 in 20 to 100% of the wells. By this time surviving spleen cells will have organized into collections which may grow strongly and threaten the survival of slow-growing hybridomas. A successful fusion requires not only good fusion conditions and a rich growth-promoting medium but restraint of growth of the background spleen cells. These may be macrophages, fibroblasts, mast cell precursors (12) and lymphocytes which become organized into 'balls' or 'bunches of grapes'. The vigour of these background cells depends on their state of activity within the spleen at the time the mouse is killed. It is enhanced by many adjuvants. One reason for adding the cells post-fusion to cold, alkaline horse serum is that this seems to depress the growth of the background cells as well as providing an extra growth stimulus to the hybridomas. The competition for growth is important only at the very early stages. Once the hybridomas have become established they outgrow and overcome all other cell types present.

### 4.2 Protocol B

Again it is convenient to perform two fusions on the same day using the spleens from two immunized mice. We usually run the fusion experiments sequentially in a morning or afternoon.

The following preparations should be made immediately before fusion (based on one fusion).

(i) Sample the plasmacytoma line as described in protocol A and pipette $10^7$ viable cells into a universal container. Leave the tube at 37°C until the spleen cell suspension has been prepared.

(ii) 6 × 20 ml of RPMI-1640 medium in universal containers. Place in a water bath at 37°C or a 37°C non-gassed incubator to equilibrate to 37°C.

(iii) 4 × 24 ml of RPMI-1640 medium supplemented with 20% FCS and penicillin/streptomycin in universal containers. Keep at 37°C with the tops closed firmly so that the medium does not become alkaline.

(iv) 2 × 10 ml syringes, each containing 10 ml of RPMI-1640 medium and fitted with 26-gauge needles. Store with a shield inside the sterile package in the sterile hood.

(v) 1 ml syringe containing 0.8 ml of 50% PEG prepared in RPMI-1640 medium. Store at 37°C.

(vi) On the bench: stopwatch, 70% ethanol, absorbent paper, balance tube containing 21 ml of water, 37°C bath, sterile instruments, Petri dish containing 1 ml of RPMI-1640 medium.

(vii) In the hood: pipettes, Petri dish.

(viii) 2 × 24-well Costar 2 ml culture trays. These are used straight from stock.

#### 4.2.1 *Fusion procedure B (based on one fusion)*

Prepare the spleen cell suspension as described in protocol A. Pipette $10^8$ viable spleen cells into a plastic disposable universal container. Remove the tube of plasmacytoma cells from the 37°C incubator. Wash the spleen and plasmacytoma cells twice, in separate containers, using the RPMI-1640 medium which has been kept at 37°C. Centrifuge the cells at 200 *g* for 10 min and flick the tubes to disperse cell deposits during the washing procedures. After two washes resuspend each cell type in 10 ml of warm RPMI-1640 medium. Pour the spleen cell suspension into the plasmacytoma container and mix the cells thoroughly by inversion. Centrifuge at 440 *g* for 7 min. Use a Pasteur pipette to remove the supernatant completely leaving only the cell pellet. Flick the container to disperse the cells and proceed as follows.

(i) Place the container at 37°C, in a water bath in the hood or a non-gassed incubator, and leave for 5 min to equilibrate.
(ii) Remove the PEG syringe from the incubator to the hood. Remove the tube from the 37°C incubator and flick the tube to mix the cells and disperse the cell deposit over the conical base of the container.
(iii) Add the 0.8 ml of PEG to the tube over 1 min. Rotate the tube at an angle to ensure that the cell deposit and the PEG are gently dispersed over the base of the tube.
(iv) Allow the tube to stand at 37°C, in a water bath or non-gassed incubator, for 1 min.
(v) Over 1 min add, in drops, 1 ml of RPMI-1640 medium which has been kept at 37°C.
(vi) Over the next 5 min, slowly add 20 ml of warm RPMI-1640 medium.
(vii) Centrifuge the cells at 440 *g* for 15 min using the 21 ml balance as companion.
(viii) Pour off the supernatant and pour 24 ml of warm RPMI-1640 medium supplemented with 20% FCS and antibiotics onto the cell pellet. Replace the cap and invert the tube to resuspend and disperse the pelleted cells.
(ix) Set out and label 2 × 24-well culture trays. Using a graduated pipette or 1 ml dispenser fitted with a sterile disposable tip add 0.5 ml of the cell suspension to each of 48 wells.
(x) Add a further 1.5 ml of warm RPMI-1640 medium supplemented with 20% FCS and antibiotics to each well. Transfer the plates to a gassed incubator which is humid.

The day on which the fusion experiment is performed is counted as day 0. On days 1, 2 and 3 post-fusion, carefully replace 1 ml of medium from each well with 1 ml of HAT medium. It is important not to disturb the cells in the wells and, particularly when adding fresh medium, to avoid ejecting it such that the cells are washed to one side of the well. The reasons for not disturbing the cells are 2-fold. First, this allows an even growth of background cells essential to promote the growth of hybridomas. Secondly at later stages it is possible to see whether one or more hybridoma colonies are growing in the wells.

It is also vital to avoid cross-contaminating wells with cells. We use a 1 ml dispensing pipette fitted with a fresh disposable tip to remove medium from each of the wells. When hybridoma cells are growing vigorously in wells, changing tips for each well also ensures that supernatants taken for analysis are not contaminated with antibody from another well. Addition of 1 ml of fresh medium to all the wells can be accomplished using a single tip provided the tip is not allowed to come into contact with the wells. After changing the culture medium on days 1, 2 and 3 it is sufficient and convenient to replace 1 ml of medium with 1 ml of fresh HAT medium on Mondays, Wednesdays and Fridays. The background antibody produced by residual spleen cells is eliminated by this process of changing the medium every 2–3 days.

Trays should be inspected before changing the medium. The appearance of culture wells is as described in protocol A. Once hybridoma cells have started to grow in the wells, if the wells have not been disturbed when changing the medium, they should grow as reasonably discrete colonies. If there is growth in a large percentage of the wells take note of the number of individual colonies in each well. If the well contains more than one colony it will be essential to clone lines which have been selected during screening as soon as possible. In general, hybridomas grown in Costar wells should be cloned immediately since it is more likely in this case than when using microtitre wells that growing wells do not represent a single cell clone. A useful guide is if there is growth in 1/3 of the wells or less most of the growing wells represent clonal growth.

The supernatants are usually assayed for antibody activity between days 18 and 25. During this time the hybridoma cells are almost confluent in the wells and the medium is clearly yellow. This ensures that the supernatant contains an optimal amount of antibody for screening analyses. It is advantageous to screen supernatants from all 48 fusion wells, including non-growing wells, as this gives a positive selection over background residual antibody. Sequential analyses of antibody activity in supernatants removed at 2–3 day intervals is useful in showing whether antibody production is increasing or declining in a particular culture. The latter case may relate either to an increased residual background problem in a particular well or a hybridoma line may be being outgrown by non-secreting hybrid cells. It is difficult to rescue antibody-secreting hybridomas from growing wells in which the supernatants show a progressive decline in antibody activity. Immediate single cell cloning is recommended but may not work.

### 4.3 **Comments on various stages of the fusion procedure**

(i) It is important that the plasmacytoma should be growing well in log phase. The cells should not be allowed to overgrow the medium. It is preferable to change the medium on the day prior to the fusion. Two or three medium-sized flasks of plasmacytoma cells can be seeded at slightly different densities if you are concerned that the cells may overgrow by the next day. In this case you can choose the most suitable culture prior to the fusion. Viability, as judged by dye exclusion, should be high. Cell suspensions are normally prepared immediately prior to the fusion but the cells may be left in H-RPMI–2% FCS at 4°C for up to 2 h.

(ii) The spleen cells are best obtained by blowing them out of the spleen rather than by tearing the spleen apart or squashing it in a homogenizer because clean single cell suspensions of high viability are obtained. They retain excellent viability in the cold in H-RPMI–2% FCS. Hence it is possible and more convenient to prepare two spleens consecutively. Occasionally a very abnormal spleen with many granulocytes or abnormal mononuclear cells is encountered which may not give rise to a successful fusion.

(iii) RPMI-1640 medium used for washing the cells may be replaced by Dulbecco's.

#### 4.3.1 *Comments on procedure A*

(i) Trial tests have shown that suspension of the cells in alkaline RPMI-1640 medium (or Dulbecco's) for the fusion stage is a crucial step. For some reason the fusion occurs better at alkaline pH and the cells are not adversely affected by this brief exposure. Wrapping the cells around the wall of the tube also improves the success rate of fusions, presumably because it ensures intimate mixing and close contact of cells. It is not essential to use a 37°C bath in the hood but the cells should be kept at approximately 37°C.

(ii) Most authors recommend that the PEG be added slowly. Trial tests have shown that there is no advantage in doing this in the case of protocol A. However since the PEG is shot to the bottom of the tube and rolled around over the cells on the wall of the tube, mixing is effectively slow. It is vital nevertheless that the PEG is diluted slowly post-fusion. Several aliquots of medium added at approximately 1 min intervals meets this requirement adequately. Rapid dilution produces osmotic damage.

(iii) It is usual to add one drop of the re-suspended cell mixture post-fusion to empty wells and to keep the suspension warm. We have found some benefit in adding one drop of cold 50% horse serum to wells before adding the cells. This appears to depress the growth of the 'background' cells while not affecting the growth of the hybridomas.

### 4.3.2 *Comments on procedure B*

(i) In this procedure alkaline RPMI-1640 medium or Dulbecco's medium was not used. An alkaline pH is achieved prior to the fusion stage by washing cells in serum-free RPMI-1640 medium and then the cells, in a very small volume of medium, are left for 5 min to equilibrate to 37°C during which the medium becomes alkaline.

(ii) Again it is vital that the PEG is diluted slowly post-fusion. The colour of the cell pellet when the cells are centrifuged post-fusion gives an indication as to whether osmotic damage has occurred. If the cell pellet is considerably paler in colour than that of the original mixed spleen and plasmacytoma cells then some red cell lysis has occurred and, therefore, osmotic damage to hybrid cells. If the cell pellet is whitish in appearance the fusion is unlikely to be successful. In this case it is not worthwhile plating the cell suspension into wells.

## 5. SELECTION OF HYBRIDOMAS

The selection of hybridomas which produce antibodies of interest usually involves a two-stage screening strategy. Culture supernatants are first assayed for activity against the immunogen. Positive supernatants are then screened against a test panel of antigens to reveal whether the antibody binds selectively to the antigen of interest. Interesting hybridoma clones invariably represent only a small percentage of the number of growing or antibody-producing wells. It is important to identify favoured hybridoma clones as quickly as possible, otherwise time and materials are wasted if a large number of unwanted lines are propagated. Success in using the hybridoma approach to produce useful antibody reagents is very dependent on careful screening so as not to miss antibodies of interest and to discard unwanted clones with speed and confidence.

What form of screening assay to use should relate to the eventual use of the Mab reagents. In other words, if you intend to produce Mabs which you will want to use together with complement to lyse cells then a complement cell lysis assay should be used to screen fusions. Similarly, if Mabs are to be developed for use in precipitation assays or to stain paraffin-embedded tissue sections in immunohistological studies then these assays should be used accordingly to screen fusions.

The screening assay should ideally satisfy all the following criteria.

(i) The assay should be simple such that it can be performed successfully and conveniently by any available technical assistance in the laboratory. When a simple assay system is used it is more informative to sample the entire culture tray and thus select positive clones against background antibody produced by surviving spleen cells.

(ii) The assay should be automated as much as possible and thus can be used to screen rapidly a large number of samples from fusion wells. This can be adequately achieved by using a multichannel pipette (eight samples at once) and by performing assays in trays.

(iii) The assay should be capable of detecting concentrations of antibody of less than $1-10$ $\mu$g/ml. Positive values in relation to non-specific backgrounds in assays should be over three times the background.

(iv) Results from screening assays should be available by at least the following day so that immediate attention can be given to favoured clones to prevent overgrowth.

(v) Reliability is important and appropriate positive and negative controls should be included in screening assays.

(vi) If a second-stage fluorescein, peroxidase or $^{125}$I-labelled antibody to mouse immunoglobulin is used this should be carefully selected to detect all classes of mouse immunoglobulin ($IgG_1$, $IgG_{2a}$, $IgG_{2b}$, IgG3, IgA and IgM).

(vii) An inexpensive screening assay is recommended in view of the requirement for screening large numbers of samples.

Remember to take supernatants for analysis from wells in which good growth of hybridomas has occurred; in this way the sample will contain an optimal amount of antibody. A very simple radioimmunoassay can be used to check that adequate levels of antibody are present in supernatants and routinely monitor immunoglobulin production in culture wells (13). It will be well worth doing this if you have a large number of growing wells and yet see no positive results when the supernatants are tested in the screening assay you have chosen to use. If you find good antibody titres in the assay for mouse immunoglobulin then the sensitivity of the primary screening assay is inadequate and should be improved. If you find low amounts of mouse immunoglobulin, then allow the hybridoma cells to grow further and take fresh samples for screening. Alternatively, you may concentrate the supernatants for analysis, although this should not normally be necessary.

In view of the above considerations, the assays you have routinely used in the laboratory with high titre polyclonal antisera may have to be adapted for screening hybridomas. Some assays may be too complicated; for example, you may wish to produce Mabs which inhibit function in a biological assay. In using complicated and lengthy assays it is advantageous to reduce the number of samples for analysis. A simple screening assay for mouse immunoglobulin can be used to identify antibody-producing clones and thus only these samples, together with a negative control supernatant, are assayed in the more complicated assay. We have found, in producing Mabs to cell surface antigens, that approximately half of the growing wells produce mouse immunoglobulin and again approximately half or more of the antibody-containing supernatants show reactivity against the immunizing cells.

### 5.1 Screening assays for antibodies to cell surface and other particulate antigens

Some form of binding assay is invariably used. The target cells may be live, glutaraldehyde-fixed, dried on a slide or be present in a frozen histological section. The test is inevitably indirect and the second-stage reagent may be fluorescein-, enzyme or radioiodine-labelled sheep, goat or rabbit anti-mouse immunoglobulin. Typically, a sample (one drop) of supernatant is added to $0.5 \times 10^6$ target cells in a small volume in a small plastic tube or tray. After 30–60 min incubation at 4°C the suspension is diluted, the cells spun down, washed once and exposed to a suitably diluted polyvalent fluorescein-conjugated anti-mouse immunoglobulin for 30 min at 4°C. The cells are washed again and the fluorescence read directly under a fluorescence microscope or by putting the cells through a fluorescence-activated cell sorter (FACS) (14). Detailed procedures for the above assays are described in Volume II.

### 5.2 Screening assays for antibodies to soluble antigens

The main alternatives are enzyme-linked immunosorbent assays (ELISA), radiobinding or haemagglutination assays. In the first two techniques the wells of a plastic microtitre tray are coated with antigen (for 1 h), washed and, after exposure to one drop of supernatant for 30 min or overnight, the wells are washed and exposed to enzyme or radioiodine-conjugated anti-mouse immunoglobulin for 1 h. This is followed by further washing and either development of enzyme activity (ELISA) or cutting out the wells and counting the $\gamma$ emissions (radioassay). Detailed procedures are described in Volume II.

For passive haemagglutination, antigens are coupled to red cells by the chromic chloride technique and stored sterile until required (15). The test, which is described in detail in Chapter 7, simply consists of adding one drop of supernatant to one drop of 0.5% coated sheep red cells. The cells are allowed to settle for 1–2 h at room temperature before reading the settled pattern.

If fusion procedure A is used and hybridoma cells are grown in microtitre wells the entire tray can be sampled using a multichannel pipette. The samples removed are transferred directly to a round-bottomed microtitre tray. The sample trays are collected and, when complete, each well receives one drop of a 0.5% suspension of coated sheep red cells. A competing agent can be added in the diluent to test the specificity of Mabs. For example, if it is desired to pick up Mabs to human $IgG_2$, $IgG_2$-coated red cells are used and $IgG_1$ is added to the diluent. In this case only Mabs which specifically recognize $IgG_2$ will agglutinate the coated red cells and Mabs which bind to antigens shared between $IgG_2$ and $IgG_1$ bind mostly to the excess of $IgG_1$ in the diluent. Favoured clones are grown up in culture as described in Section 6.

## 6. CLONING OF HYBRIDOMA CELL LINES

Cloning is essential to confirm monoclonality of the antibody preparation. Whether cloning is undertaken immediately or at a later stage depends somewhat on whether the primary hybridoma cultures were grown in microtitre wells (fusion procedure A) or in 2 ml wells in a 24-well culture tray (procedure B).

### 6.1 **Fusion procedure A**

When fusion experiments are plated into microtitre wells, cells producing antibody of a desired specificity are transferred to 2 ml culture wells of a 24-well culture tray. When confluent growth is reached in the 2 ml culture well and the medium has turned yellow the supernatant may be re-screened against a suitable panel of antigens. If the test results are favourable, the hybridoma is now grown up in several wells and then gradually expanded in universal containers and then flasks of increasing size. By this time the cells have usually cloned themselves, the best grower (hopefully the antibody-producer) having taken over the culture.

Cloning is best done either at a very early stage to attempt to rescue a slow growing clone of cells from fast growing neighbours or at a final stage when a stable antibody-producing line has been achieved. In the latter case cloning is then usually nothing more than a formal confirmation of monoclonality. Even if cloning is avoided at the production stage it must be undertaken later, even with lines already cloned, in order to eliminate non-producer variants which sometimes appear after long-term growth of clones. These may gradually take over the culture if they are not cloned out. Some lines are extremely stable and never need re-cloning. Others alter their growth characteristics and antibody production levels. Sometimes the situation can be rescued by re-cloning but some lines are inherently unstable. As soon as a line has been cloned, immediately store at least 10 vials of cells in liquid nitrogen which you can return to if on prolonged growth the cloned line is unstable.

The more successful the fusion experiment, the more likely it is that more than one clone will grow up in individual wells and that two clones of similar growth rates will persist. Mixed clones may be detected by the presence of more than one specificity or antibodies of more than one class or subclass and by complex spectrotypes on iso-electric focusing. In this case, immediate cloning is essential.

### 6.2 **Fusion procedure B**

In this case, when fused cells are plated into 2 ml culture wells, there is an increased likelihood of more than one clone within a single well. You may have observed two or more clones in wells when cultures were inspected under the inverted microscope. Cloning of cells producing antibody of a desired specificity is undertaken immediately using cells from the primary culture well, when confluent (usually 21 – 28 days after the initial fusion). Nevertheless, the hybridoma culture is still expanded in several wells and then in flasks of increasing size so as to store, in liquid nitrogen, at least five vials of the original cell population if the cloning is difficult or initially unsuccessful. As the hybridoma cells are grown in flasks the culture supernatant is monitored for antibody production. A progressive decline in antibody production when the culture has reached the flask stage indicates that the antibody-producing cells are being outgrown by non-secreting variants. It is then essential to isolate the antibody-producing hybrid cells in the cloning experiments and there is little value in storing cells from these flasks for further attempts at cloning. Attempts to clone antibody-secreting hybrid cells from flasks in which the supernatant shows a continual loss of antibody activity are invariably unsuccessful. If it is vital to isolate the line, recover the earliest stored vial of cells from liquid nitrogen and clone as soon as cells have re-established themselves.

Cloning may be performed either by seeding a small inoculum of cells for growth in soft agar or by limiting dilution.

### 6.3 **Cloning of cells in semi-solid agar**

For this cloning procedure the following materials are required:

(i) Hybridoma cells at $3 \times 10^5$/ml in complete HAT medium.

(ii) HAT medium supplemented with 50 $\mu$M 2-mercaptoethanol, kept at 44°C in a water bath.

(iii) 3% and 5% Bacto-Agar prepared in 0.9% saline. The agar is autoclaved in small aliquots in glass universal containers and can be stored at 4°C.

(iv) 5 cm diameter tissue culture grade Petri dishes.

The cloning procedure is given in *Table 2*.

### 6.4 **Cloning of cells by limiting dilution**

When hybridoma cells grown in 2 ml culture wells, as in fusion procedure B, were cloned in soft agar, as described above, in some instances a large number of colonies had to be picked before an antibody-secreting colony was isolated. For this reason,

**Table 2.** Steps in the procedure for the cloning of cells in semi-solid agar.

| | |
|---|---|
| 1. | Melt agars at 100°C in a pan of boiling water and transfer to a water bath at 44–45°C. Take care when handling the universal of hot agar. |
| 2. | Dilute the 5% agar stock to 0.5% (1:10) with the HAT medium, mix well and keep at 44°C in a water bath. It is convenient to prepare three bases at a time (1.5 ml of agar + 13.5 ml of medium). |
| 3. | Pipette 9 × 5 ml base layers of 0.5% agar evenly over the surface of 5 cm Petri dishes. Allow the bases to set. |
| 4. | Using the 3% agar, prepare 9 × 5 ml aliquots of 0.3% agar in HAT medium and keep the tubes in the 44°C bath. |
| 5. | Immediately before pouring onto the base layer, add one, two or three drops of the hybridoma cell suspension to an aliquot of 0.3% agar, mix well and pour evenly over the 0.5% agar base. Prepare three plates for each number of drops. |
| 6. | Stack the plates in a clear plastic sandwich box which contains a Petri dish filled with sterile distilled water. This ensures a humid atmosphere in the box. Transfer the box with the lid loose to the $CO_2$ incubator which should be humid. Allow the box and plates to equilibrate with the 5% $CO_2$ at atmosphere, replace the lid and leave for 10–14 days. |
| 7. | Pick off colonies at 10–14 days using Pasteur pipettes and transfer to microtitre wells. |
| 8. | Test the microtitre well supernatants for antibody activity when confluent. At this time remove all the medium and replace with fresh HAT medium. |
| 9. | Expand the desired positive clones first in microtitre wells, then to 2 ml wells and finally to small flasks. |
| 10. | Freeze down at least 10 vials of the cloned line in liquid nitrogen. |

cloning was carried out in 96-well flat-bottomed mitrotitre plates using a series of cell dilutions. This approach has the advantage that antibody production by single colonies growing in wells seeded with one cell per well can be readily monitored, in the culture supernatant, throughout the cloning procedure. The limiting dilution procedure is easy to perform and avoids the problem with the agar technique that success is often dependent on the batch of agar and consistency.

The limiting dilution method is a three-step procedure and does take longer to derive cloned cell lines.

### 6.4.1 *Primary dilution series*

Throughout the cloning, hybridoma cells can be selected for good growth and high viability and for high levels of antibody production. The aim of the primary dilution series is to select a sub-culture from the original 2 ml culture cell population which secretes antibody at higher levels than the original well.

(i) Plate the hybridoma cells, taken from the original 2 ml culture, into the microtitre plate at a density of 100 cells/well ($5 \times 10^2$ cells/ml) in 200 $\mu$l of HAT medium. Use the inner 48 wells of the plate (6 rows down $\times$ 8 wells across) and fill the outer wells with RPMI-1640 medium containing antibiotics to prevent dehydration and concentration of the HAT medium in the wells containing cells.
(ii) Incubate the plate in a $CO_2$-gassed incubator which is humid.
(iii) After a period of around 7 days, when growth is almost confluent, assay the supernatants for antibody activity.

It is advisable to test the neat supernatant and one or two dilutions (1/4 and 1/16) if possible. Thus wells showing increased antibody titres in comparison with each other and containing viable cells which are growing well are selected for single cell cloning experiments. Two wells are selected and cells plated in the following secondary dilution series.

### 6.4.2 *Secondary dilution series*

This aims to select hybrid cell lines grown from a single cell and requires the addition of a spleen cell feeder layer.

(i) Prepare the spleen cells, as described for fusions, the day before performing the cloning experiment and re-suspend the cells at $10^6$/ml in HAT medium supplemented with 100 $\mu$M 2-mercaptoethanol.
(ii) Leave the spleen cells, in a universal container, overnight at room temperature in the sterile hood. This procedure has the advantage that the next day you can check on the microscope that the spleen cells are not contaminated; it probably also depresses the growth of background spleen cells in the cloning wells.
(iii) Prepare two microtitre plates as follows. Again use only the inner 48 wells of the plate (6 rows $\times$ 8 wells) and fill the outer wells with RPMI-1640 medium containing antibiotics. To row 1 (8 wells across) add 100 $\mu$l of complete HAT medium. To rows 2–6 down (each 8 wells across) add 100 $\mu$l of the spleen cell suspension to each well ($10^5$ cells/well).
(iv) Place both the microtitre plates in the $CO_2$-gassed incubator until you have pre-

pared the dilution series of the hybridoma cells. It is important to prepare the dilution series of hybridoma cells and plate them as quickly as possible. If cells are left for long at low cell numbers they will lose their viability. Two secondary dilution plates are prepared using cells from two wells of the primary plate. The dilution series of the two hybridoma cell populations are prepared sequentially and cells are removed from the wells of the primary plate as required.

(v) Prepare the following volumes and cell concentrations in complete HAT medium: 1 ml of cells at $10^3$ cells/ml, 1 ml of $10^2$ cells/ml, 1 ml of 50 cells/ml, 2 ml at 10 cells/ml and 1 ml at 5 cells/ml. Note that cell dilutions are prepared as serial dilutions so that you will have to start with approximately 1.5 ml of cells at $10^3$ cells/ml.

(vi) Remove one of the prepared microtitre plates from the incubator. To row 1, which contains only HAT medium, add 100 $\mu$l of cells at $10^3$/ml to each of the eight wells. These wells, now containing 100 cells/well, ensure a continuous stock of the cell line selected from the primary dilution plate. To row 2, add 100 $\mu$l of cells at $10^2$/ml to give 10 cells per well. Row 2 and all the subsequent rows contain the spleen feeder cells. To row 3, add 100 $\mu$l of cells at 50 cells/ml (5 cells/well), to rows 4 and 5 add 100 $\mu$l of cells at 10 cells/ml (1 cell/well) and to row 6 add 100 $\mu$l of cells at 5 cells/ml (1 cell every two wells). The final concentration of 2-mercaptoethanol in the wells now containing spleen cells and hybrid cells is 50 $\mu$M.

(vii) Transfer this plate to the $CO_2$-gassed incubator and repeat the procedure for the second well of cells selected from the primary dilution plate.

(viii) Change the medium in the wells if they become too acid whilst the colonies are growing. Take care not to disturb the cells in the wells as hybrid cells will then grow as discrete colonies on the spleen cell layer. Avoid cross-contaminating the wells when changing medium.

(ix) After 10−14 days the hybrid clones should be visible to the naked eye, and clearly visible using the inverted microscope and a low power objective, as discrete white spots in the wells. Record the number of colonies growing in each well. Ideally colonies grown in wells seeded with 1 or 0.5 cell per well should be selected. However, if the plating efficiency of the line is low you may have to choose a clone from a well seeded at 10 or 5 cells per well provided only a single colony is present in the well.

(x) When the colonies are quite clearly visible to the naked eye they are usually large enough to be assayed for antibody production. At this stage assay supernatants from the complete plate for antibody activity. Again, if possible, assay the neat supernatant and one or two dilutions. Select three individual colonies which produce a high titre of antibody and preferably colonies from rows in which cells were plated originally at 1 or 0.5 cell per well.

### 6.4.3 *Third dilution series*

To ensure that these clones are derived from a single cell and not two cells it is advantageous to repeat the above secondary dilution experiment for each of the three colonies selected. Repeating the single cell cloning experiment also has the advantage that when individual colonies derived from the third dilution series are assayed they should all produce antibody. Thus the hybridoma line which is finally grown to large volumes is with certainty derived from a single cell. It is worthwhile plating three separate colonies from the secondary plates to allow for the possibility of poor plating efficiency of one or two of the colonies. By transferring three clones you are certain to grow enough single cell colonies in the three tertiary plates for analysis.

The single cell colonies grown in the third dilution series and assayed for antibody production are transferred to fresh microtitre wells. Eventually one preferred line is expanded first to microtitre wells, then to 2 ml wells and finally to small flasks. At least 10 vials of cells are stored in liquid nitrogen.

#### 6.4.4 *Comments on the dilution procedure*

At first sight, the above procedure appears to involve a lot of work. The strategy is particularly useful if one or two dilutions of the supernatant from each well are assayed and the titre of antibody can then be assessed. Thus cells producing increasing amounts of antibody can be selected throughout the cloning procedure. In our own laboratory, by using a multichannel pipette (eight samples at once) to sample hybrid supernatants, preparing dilutions of supernatants in non-sterile microtitre plates and then transferring these dilutions to plate radioimmune assays it is possible to assay the 48 samples from each cloning plate very quickly. With practice the limiting dilution plates of hybridoma cells can be set up in a short time.

It is important to keep good records of cell cultures and antibody activities in culture supernatants during the growth of fusion wells and throughout cloning experiments.

## 7. HARVESTING OF ANTIBODIES IN CULTURE AND AS ASCITIC FLUID

There are two sources of Mabs from hybridomas.

### 7.1 **Culture supernatant**

This contains no mouse immunoglobulin other than the Mab, but of course does contain large amounts of FCS proteins. Tissue culture supernatants contain $\mu$g/ml levels of antibody. Modern serum-free culture media are increasingly being used to prepare pure Mabs.

To obtain optimal amounts of antibody in supernatants, grow the hybridoma cells to a high density such that the medium becomes yellow. However, it is important not to allow the cells to overgrow to an extent that the cells become unhealthy and start to die rapidly. Hybridoma cultures containing a high percentage of non-viable cells can often be difficult to re-establish and may often have to be discarded. The supernatant is harvested, centrifuged at 500 *g* for 5 min to remove cells. The cell-free supernatants from multiple flasks of one hybridoma line are pooled in clean glass bottles. To provide supernatant for one's own use and to supply other investigators we recommend that you prepare a 500 ml batch of supernatant which is stored at 4°C containing 0.1% sodium azide. Most antibodies are stable and this batch of supernatant, which is checked for activity, can be used for years.

7.2 **Ascitic fluid from a mouse injected intraperitoneally with live hybridoma cells**

(i) Prime BALB/c mice, 3–4 months old, by i.p. injection of 0.5 ml of Pristane oil (2,6,10,14-tetramethyl pentadecane; Koch-Light) 1–3 weeks prior to injecting hybridoma cells grown in tissue culture.

(ii) Inject each mouse with at least $10^6$ and, preferably $10^7$, cells in PBS. The hybridoma grows as an ascitic tumour and the ascitic fluid contains antibody at approximately 10–500 times the concentration in culture supernatant (mg/ml levels of antibody).

(iii) To harvest the ascitic fluid, kill the mouse by breaking the neck, make an incision to open the peritoneum to allow access with a Pasteur pipette. Drain the peritoneum of ascitic fluid.

(iv) Centrifuge the fluid at a high speed to remove cells and debris. Pooled ascitic fluid is best stored at −20°C in 1 ml aliquots or as a large batch if the antibody is to be further purified.

Ascitic fluid is, of course, contaminated with other mouse immunoglobulins, the level depending on the cleanliness of the animal house. Colonies of SPF mice are preferable if available. The level of contaminating antibodies is also raised if there is much effusion of blood into the ascitic fluid. Some hybridomas grow well and diffusely in ascitic fluid. Occasional lines produce adequate amounts of antibody in tissue culture and, despite growing well as ascites, produce little antibody *in vivo*. Other hybridomas form solid tumours or metastasize or occasionally kill the host before much growth has occurred.

## 8.QUALITY CONTROL: CELL LINES AND ANTIBODY PREPARATIONS

The first essential in quality control is to have frozen stocks of the hybridoma cell lines date-recorded. Samples of cultured cells should be frozen down from each phase of growth and ascitic cells from every few passages in mice should be sampled and frozen with an indication of the passage number.

8.1 **Freezing down cells**

A simple effective freezing-down technique is as follows.

(i) Use cells from a rapidly dividing healthy culture. Spin down in universal containers approximately $10^7$ cells for each vial to be frozen.

(ii) Remove the supernatant and resuspend the cells at $1 \times 10^7$/ml in 1 ml of cold 5% dimethyl sulphoxide in 95% FCS. The freezing mixture should be made up previously and stored at −20°C in aliquots.

(iii) Transfer the cells to small polypropylene freezing ampoules which are labelled, using an indelible pen, with the line code and data.

(iv) Place the freezing vials into an expanded polystyrene box, with walls about 2–3 cm thick and a lid, and transfer the box to a −20°C freezer and leave for 30 min. Suitable containers for freezing are the small boxes in which breakable laboratory materials are supplied. The expanded polystyrene container ensures that the vials of cells are cooled slowly (ideally at 1°C per min) down to −20°C.

(v) After 30 min, transfer the box to −70°C and leave either for 6 h or overnight before removing the vials and transferring to a liquid nitrogen storage container.

An alternative freezing protocol is to transfer the polypropylene tube to a 50% glycerol bath set at −32°C. After 40 min remove the tubes, wipe with tissue and transfer directly to the liquid nitrogen container.

### 8.2 **Recovery of frozen cells**

(i) To recover cells thaw the tube quickly by holding it, with light agitation, in a stream of hot water from the tap until the ice has almost disappeared.
(ii) Transfer the cells to a universal container and then slowly add, in drops, 2.0 ml of warm (37°C) H-RPMI.
(iii) Replace the cap and mix. Centrifuge for 5 min at approximately 500 *g*.
(iv) Aspirate and discard the supernatant and take up the cells in 4 ml of HAT medium and transfer to 2 × 2 ml culture wells.
(v) Incubate in a gassed incubator and expand the culture when appropriate. Recovery of viable cells is normally 50−78% provided that healthy vigorously growing (not overgrown) cells are frozen down.

From time to time the specific antibody content of culture supernatants must be checked by titrating supernatants against a limited panel of test antigens. It is important to check not only the antibody level but for possible contamination with other Mabs. These might be present if the hybridoma line itself has been contaminated with another line which can easily happen when many lines are being grown up simultaneously.

Cells from ascites are similarly cryo-preserved but in this case there is no need to observe sterile precautions which must be observed when freezing cells for *in vitro* culture. Cells obtained directly from a mouse are the source material for future mouse passages whereas frozen cultured cells are used for *in vitro* culture. Recovered ascites cells for injection into mice should be taken up in PBS (not HAT medium).

The simplest way of monitoring the level of the monoclonal immunoglobulin in ascitic fluids is to run an electrophoretic strip and stain it conventionally. A single band in a characteristic position which is different for each Mab should be seen. Its intensity may be compared with fluids from previous passages. This is more convenient and informative than titrating the specific antibody content although this must also be checked from time to time.

## 9. APPLICATION OF MONOCLONAL ANTIBODIES

The clinical and experimental applications of Mabs are varied and numerous. A number of reviews and chapters in books have been written entirely on this topic and are recommended (2,4,6). Here we consider practical issues relating to the use of Mabs.

First, there is a choice as to suitability and convenience of tissue culture supernatant or ascitic fluid as a source of antibody in various applications. The important consideration is whether admixture of Mab with naturally occurring murine antibodies, in the case of ascitic fluid, will lead to difficulties in interpreting results and/or unacceptably high backgrounds in assays. Clearly ascitic fluid contains mg/ml amounts of Mab and when used dilute (at 1 in $10^3$ or greater) most irrelevant murine antibodies will be diluted out. However, in some cases the presence of contaminating murine antibodies should be avoided entirely. For example, if Mabs are used to isolate a small amount of protein antigen in an extremely pure form for precise biochemical analyses, tissue culture supernatant is a preferable source of antibody. This avoids spurious contamination of the required antigen.

Whether ascitic fluid gives unavoidable background problems in the assay you use will have to be considered. In some instances, mice may have high titre natural antibodies which bind to the antigen specifically recognized by the Mab. There are also non-specific problems in using ascitic fluid. For example, tissue culture supernatant gives a clearer and more precise staining pattern in immunohistological studies using Mabs.

Tissue culture supernatant provides a convenient and adequate source of antibody for use in antigen detection tests in one's own laboratory and to supply colleagues with reagent. Used at one drop of neat supernatant per test, 10 ml is sufficient for approximately 200 tests. If Mabs are to be provided commercially or for use in large studies in a number of centres, ascitic fluid is a more economical source of antibody. Used at one drop per test and at a 1 in $10^3$ dilution of ascitic fluid, 1 ml provides 20 000 tests. In this case, the Mab is usually isolated as an immunoglobulin fraction from the ascitic fluid.

In Section 5 we recommended that the choice of screening assay should relate to the eventual use of Mabs. A consequence from this statement is that Mabs may not reveal antibody when used in other assay systems. For example, Mabs to soluble proteins selected by passive haemagglutination using antigen coupled to red cells may recognize a single, non-repetitive determinant. As such these Mabs, unlike conventional polyclonal antisera, will not precipitate antigen when used individually. Antigen precipitation can be achieved using a combination of two Mabs which identify two different determinants on the protein.

Problems in undertaking further studies using Mabs are encountered when the Mab identifies the antigen only in a native form. For example, Mabs to cellular antigens often do not reveal antigen on nitrocellulose blots prepared from SDS – polyacrylamide gels of soluble cell extracts. In this case the antigen is denatured during these processes and thus does not bind antibody. Similarly, Mabs which stain cells in suspension may or may not be suitable for immunohistological studies using paraffin sections of formalin material. These considerations suggest that the choice of screening assay and nature of material used to immunize mice for fusion experiments are both important factors influencing the properties of the Mabs finally selected.

The serology of Mabs cannot always be predicted from known serological patterns obtained using absorbed polyclonal antisera. In the case of polyclonal reagents, the pattern of reactivity represents the combined effect of a number of different antibodies which together identify one cell type or protein and not others. Mabs raised to particulate antigens or glycoproteins may show unexpected serological patterns. They may stain various unrelated cells or bind to many different glycoproteins. These antibodies may identify specific carbohydrate structures which can occur in a remarkably diverse range of proteins. This point is exemplified by the carbohydrate structure 3-fucosyl-*N*-acetyllactosamine, which can be detected in $\alpha$-1-acid glycoprotein, lactoferrin, parotid $\alpha$-amylase, cervical mucin and secretory component. Mabs which show a diverse range of reactivity may have several practical applications. Serological testing should determine that the Mab is operationally specific in a particulate test system.

In general, the use of Mabs in experimental and clinical studies eliminates the problem of non-specific reactions of conventional antisera due to irrelevant antibodies or antigen binding non-immunoglobulin components. However, as shown above, you may require more than one Mab to a particular antigen in order to undertake a wide variety of studies.

## 10. PROSPECTS FOR THE FUTURE

New specificities continue to be produced at a rapid pace and before a decade passes there are likely to be something of the order of $10^6$ Mabs derived from rat and mouse hybridomas. Most of these hybridomas will be scattered around the world in individual laboratories. Obviously there will be a great deal of repetition of some specificities within different centres. This presents a problem other than one of mere repetition. An important attribute of Mabs is that they can be produced in sufficient quantities for distribution to a large number of laboratories undertaking similar or complementary analyses. Therefore, data from these studies can be meaningfully correlated. However often different laboratories use their own Mabs which appear to identify the same antigen; that this is not certain undermines the possibility of correlating data from different centres. The results of international workshops in which a number of laboratories compare Mabs using various techniques will help to overcome this problem. For example, there has been a great outpouring of Mabs to human leukocyte antigens. Two international workshops (Paris 1983; Boston 1984) have organized the testing of Mabs to T, B and myeloid lineage antigens in cytological, histological and biochemical assay systems. These results have been very helpful. Many of the antibodies have been shown to belong to specificity clusters and have been allotted cluster-designation (CD) numbers and the target antigens partially characterized. Another recent workshop has assessed Mabs to human IgG sub-classes. Recurring workshops in these and other important fields will be most worthwhile in view of the above considerations.

Some of the Mabs so far produced are available commercially, regrettably often at very high prices. This may to some extent contribute to the repetitiveness of Mabs if centres decide it is more cost effective to produce their own reagents for large studies. An important future development is for specialist centres in the USA (American Type Collection) and in Europe to hold frozen stocks of cluster-designated hybridoma clones available to all for a handling fee. Thus, many Mabs which are at the moment prohibitively expensive will inevitably become much cheaper as alternative clones of the same specificity are deposited by altruistic research workers and the organizers of workshops. However, even if a clone is available cheaply, the cost of growing it up and storing it for future use is not inconsiderable. There is a great need for research on cheaper ways of conserving animal cells.

The therapeutic usefulness of Mab *in vivo* is an expanding area of considerable interest. For this purpose human Mabs would obviously be preferable. They may be used in diagnosis for tumour imaging or therapy to delete tumour cells. It has proved surprisingly difficult to isolate human plasmacytomas suitable for cell fusion work. The EBV-transformation technique provides an alternative means to generate human cell lines secreting Mabs. Progress to date using this technique is described in Chapter 4. A considerable amount of work is now being done in this field and will ensure a constantly changing scene. Success will reveal, in some cases, different antigens from those seen in the murine system, with development, for example, of HLA typing antibodies.

## 11. REFERENCES


1. Kohler,G. and Milstein,C. (1975) *Nature,* **256**, 495.
2. Yelton,D.E., Margulies,D.H., Diamond,B. and Scharff,M.D. (1980) In *Monoclonal Antibodies; Hybridomas — A New Dimension in Biological Analyses.* Kennett,R.H., McKearn,T.J. and Bechtol,K.B. (eds), Plenum Press, New York, p. 3.
3. Kozler,D. and Roder,J.C. (1983) *Immunol. Today,* **4**, 72.
4. Galfré,G. and Milstein,C. (1982) In *Properties of the Monoclonal Antibodies Produced by Hybridoma Technology and Their Application to the Study of Diseases.* Houba,V. and Chan,S.H. (eds), UNDP/World Bank/WHO, p. 1.
5. Kennett,R.H., Davis,K.A., Tung,A.S. and Klinman,N.R. (1978) *Curr. Top. Microbiol. Immunol.,* **81**, 77.
6. Bastin,J.M., Kirkley,J. and McMichael,A.J. (1982) In *Monoclonal Antibodies in Clinical Medicine.* McMichael,A.J. and Fabre,J.S. (eds), Academic Press, New York, p. 503.
7. Westerwoudt,R.J., Blom,J., Naipal,A.M. and Van Rood,J.J. (1983) *J. Immunol. Methods,* **62**, 59.
8. Fisher,A.G., Bunce,C.M., Toksoz,D., Stone,P.C.W. and Brown,G. (1982) *Clin. Exp. Immunol.,* **50**, 374.
9. Sarnesto,A., Ranta,S., Seppälä,I.J.T. and Mäkelä,O. (1983) *Scand. J. Immunol.,* **17**, 507.
10. Phillips,D.J., Reimer,C.B., Wells,T.W. and Black,C.M. (1980) *J. Immunol. Methods,* **3**, 315.
11. Galfré,G., Howe,S.C., Milstein,C., Butcher,G.W. and Howard,J.C. (1977) *Nature,* **266**, 550.
12. Schrader,J.W. and Nossal,G.J.V. (1980) *Immunol. Rev.,* **53**, 61.
13. Brown,G., Kourilsky,F.M., Fisher,A.G., Bastin,J. and MacLennan,I.C.M. (1981) *Hum. Lymphocyte Differ.,* **1**, 167.
14. Anderson,K.C., Park,E.K., Bates,M.P., Leonard,R.C.F., Hardy,R., Schlossman,S.F. and Nadler,L.M. (1983) *J. Immunol.,* **130**, 1132.
15. Ling,N.R., Bishops,S. and Jefferies,R. (1977) *J. Immunol. Methods,* **15**, 279.

CHAPTER 4

# Human monoclonal antibodies

JOHN GORDON

## 1. SOME GENERAL CONSIDERATIONS

### 1.1 The need for monoclonal antibodies of human origin

As detailed in Chapter 3, the production of murine monoclonal antibodies (Mabs) can be made a routine for any laboratory possessing the appropriate animal stocks and tissue culture facilities. By comparison, the ability to generate Mabs of human derivation remains problematical. The question needs to be raised, therefore, as to why any considerable investment should be made in this aspect of Mab technology. Two major areas can be highlighted where development of human Mabs is essential. One is for antibodies to human antigenic systems where the specificities demanded are too fine to permit immunological discrimination by non-primate animals (i.e. rodents). In practice, the best examples for this are found in polymorphic systems, as typified by the major histocompatibility complex (MHC) class II antigens. Despite much effort, rodent Mabs raised against human class II antigens have not been capable of providing the repertoire of specificities required for dissecting the private determinants necessary for tissue-typing. Another example where xenogeneic antibodies fail to recognize the necessary antigenic structure is in the case of the Rhesus D antigen. At present both the detection of this antigen and the treatment of the associated haemolytic disease of the newborn ('blue babies') through passive immunotherapy, relies totally upon the use of pooled human sera from appropriate donors.

The second area of need for human Mabs lies in their application for therapy or diagnosis by infusion. Cancer management is the main interest for this approach. A patient making his own antibodies to a foreign rodent Mab would lead not only to the abrogation of therapeutic effects, but also to the risk of severe clinical complications. These problems are compounded by the need for repeated large doses of antibody due to the ability of tumour cells to evade immune attack (1).

### 1.2 Limitations in the production of human monoclonal antibodies

Given the clear desirability for production of Mabs of human origin, the number that can presently be cited is remarkably small. This reflects the current limitations in generating Mabs in the human system. The problems involved warrant careful consideration as they highlight the fundamental differences in the strategies adopted for producing human and rodent Mabs.

One immediate difference lies in the potential for immunization. Protocols adopted for obtaining an aggressive antibody response in rodents could clearly not be sanctioned for use in man. In the majority of cases it is necessary to rely on fortuitous immunizations for obtaining precursor cells of appropriate antigen specificities. This limitation, at least in qualitative terms, turns out to be not as severe as might be anticipated. Many of the antigen systems where human Mabs are needed actually do generate an immune response under natural circumstances. Pregnancy, for example, gives rise to maternal antibodies recognizing the paternal antigens which form the basis of tissue and blood group typing. Tumour-directed antibodies, where they exist, could provide a cancer patient's own immune cells. Similarly, viral and bacterial infection also elicit a response from which appropriate cells could be obtained. Finally, use could be made of an individual's 'natural' repertoire of antibody specificities. Within this setting, auto-antibodies can be considered, and the availability of such human Mabs would facilitate study of the pathogenesis of the associated diseases.

The source of cells for generating Mab highlights another major difference between mouse and man. The only practical option for generating human Mabs is peripheral blood. Lymph node cells have also occasionally been used in cancer patients. The paucity of plasma cells in peripheral blood dictates that new strategies must be adopted to use the precursors of antibody-producing cells.

Success in human Mab technology has also been hampered by the low number of human myeloma partner cells which have been available for fusion. Limitations in this area are being overcome, however, through a diversity of approaches. Whilst some workers persist with the search for new human partners, others are finding increasing success using conventional mouse myelomas for fusion. In addition, the so-called heteromyelomas, constructed from mouse and human cells, have found some success and may provide the greatest compromise between stability of phenotype and high-rate antibody production.

Many of the problems encountered in adopting the somatic hybridization approach, however, are overcome in an alternative strategy, uniquely available to the human system, which is afforded through the ability of the Epstein–Barr virus (EBV) to transform and immortalize human B-lymphocytes without resort to a neoplastic partner (2,3). The application of EBV transformation to the generation of human cell lines producing Mabs represents the essential methodological difference in strategies adopted for man and mouse, and as such will be highlighted throughout the following sections.

## 2. MATERIALS REQUIRED ADDITIONAL TO THOSE DESCRIBED FOR MURINE MONOCLONAL ANTIBODIES

### 2.1 Epstein–Barr virus

A variety of sources of EBV is available, but the most widely used is that derived from the B95-8 marmoset cell line. Many institutions carry this line but in case of difficulties the author can be contacted for a supply. It is highly desirable that the line used to generate virus is completely free of mycoplasma. The cloning efficiency of B-lymphocytes infected with EBV declines drastically in the presence of mycoplasma and their long-term growth becomes problematical. Not only should the line be checked on arrival for mycoplasma but also regularly during its propagation in the laboratory.

The B95-8 line grows as a loosely adherent monolayer, when mycoplasma-free, in RPMI-1640 containing 8% fetal calf serum (FCS) plus the regular supplements of L-glutamine, antibiotics and fungicides if wished. It is normally carried in standard tissue culture flasks placed broad-side down in a moist 5% $CO_2$ incubator. Splitting and expansion of the line can be achieved by tightening the cap and swirling the culture to remove some of the attached cells for seeding into a new flask. Under normal conditions of growth this would be required 1–2 times per week.

The B95-8 line spontaneously releases virus into the medium. To harvest virus in a sufficiently concentrated form, grow the line to confluence, screw the cap down tightly and leave undisturbed in the incubator for 10–14 days. By this time the medium will have turned yellow. Decant this from the flask and pass aseptically through a 0.45 $\mu$m filter to exclude cells and large debris. Filters of lower pore size *must* be avoided as virus would inevitably be lost. If the virus is not to be used on the same day it should be stored frozen at −70°C as aliquots sufficient to infect approximately $5 \times 10^6$ cells, in practice, about 2 ml. It is worthwhile generating a large batch of virus and testing an aliquot for its potency. This is best and most rapidly done by titrating the virus preparation for its ability to induce DNA synthesis in B-lymphocytes measured by the uptake of [$^3$H]thymidine 3 days post-infection. The reader is directed to ref. 4 for this simple assay. If desired the virus can be centrifuged out of the spent medium and taken up in a smaller volume of fresh medium for storing as aliquots at −70°C. This has the dual advantage of both concentrating the virus and removing it from the rather acid environment which could prove deleterious to the cells to be infected.

### 2.2 Feeder cells

A variety of cell types has been suggested to provide good feeder layers for the cloning of EBV-transformed cells. The most commonly used have been peripheral blood mononuclear cells (PBMC) and fibroblasts, particularly human fetal lung fibroblasts. The former are more convenient to obtain and it is their preparation which is described.

There is a variety of options open to the reader for a source of PBMC. One is to use the cells remaining following the selection of cells binding the antigen of interest (see Section 3.2). Alternatively, blood from any healthy individual can be used for this purpose. It may be worthwhile to screen a number of individuals for their ability to provide a good feeder layer for EBV-transformed cells. Whoever the donor, the preparation of the PBMC is relatively straightforward.

(i) Layer whole blood diluted 1:2 with phosphate-buffered saline (PBS), by pipette onto an equal volume of a cell separation density gradient such as Ficoll-paque (Pharmacia AB, Uppsala, Sweden).
(ii) Following centrifugation for 20 min at 400 *g* collect the interface by pipette, wash three times in RPMI-1640 and re-suspend to a concentration of $2 \times 10^6$/ml.
(iii) Plate the cells in 100 $\mu$l of full medium to each well of a 96-well plate.
(iv) Irradiate the plate with 4000 rads of $\gamma$-irradiation, approximately 1–2 h prior to its use in cloning.

### 2.3 Fusion partners

There are four major choices currently available to attempt somatic cell hybridization with human antibody-producing cells. These include:

(i) human myeloma lines (5);

(ii) EBV-transformed lymphoblastoid cell lines (6);
(iii) mouse plasmacytoma lines (7);
(iv) heteromyelomas (8).

The lines used must contain the hypoxanthine−guanine phosphoribosyl transferase (HGPRT) deficiency and ideally should also carry a resistant marker such as ouabain resistance to which the partner cells will be susceptible. This is desirable as, unlike the rodent situation, the antibody partner in the human system can itself be endowed with immortality through EBV transformation. The reader is directed to the individual references for further details and will find that nearly all workers cited are willing to supply fusion partners on written request. The category of fusion partner selected is the choice of the individual experimenter but the most successful studies at present seem to have derived from using the murine plasmacytoma lines and the heteromyelomas.

## 3. DETAILED PROTOCOL FOR THE GENERATION OF HUMAN MONOCLONAL ANTIBODIES BY EBV TRANSFORMATION

### 3.1 Introduction

The nature of the antigens sought will largely dictate the precise strategies adopted, including choice of donor, preparation and enrichment of precursor cells and screening methods. Before embarking on what is likely to be an extremely laborious task, the experimenter should be convinced that not only is it both feasible and practicable to obtain the necessary precursor cells in sufficient number but that an unambiguous, simple and rapid screening method is available for the antibody specificity in question. For simplicity, the routes currently available for the production of human Mabs will be approached through considering the model system of the Rhesus D antigen. Deviations from this outline necessitated when exploring other antigenic systems will be discussed later. The Rhesus D model has been chosen as it probably represents the system studied most extensively to date and therefore allows some formative comparisons to be made and optimal procedures to be suggested.

The general principle of the strategy is to enrich, expand and clone for cells producing the desired antibody and ultimately stabilize clones through somatic hybridization.

### 3.2 Preparation of cells for EBV infection

PBMCs prepared as described in Section 2.2 consist of monocytes, T-lymphocytes and B-lymphocytes. The B-lymphocytes comprise 5−10% of total cells and of these 1% at best will provide antibodies of correct specificity. Enrichment for appropriate cells at this stage substantially reduces the volume which will need to be handled later. There are three alternatives open at this early stage.

(i) PBMCs can be infected with EBV without further separation. In this case phytohaemagglutinin needs to be added if the line is to be expanded in bulk culture, otherwise cytotoxic T-cells may be generated against EBV-transformed cells and the lines will regress (9). Alternatively, if infected cells are to be cloned immediately following infection then addition of phytohaemagglutinin is unnecessary due to effective dilution of the cytotoxic population.

(ii) T-lymphocytes can be removed prior to EBV infection by rosetting with washed sheep red blood cells. In this case, PBMC at $2 \times 10^6$/ml in RPMI-1640 containing 10% FCS are mixed with an equal volume of washed sheep red blood cells at 2% in PBS. Following incubation at 37°C for 5 min, cells are spun gently at 150 *g* for 3 min and incubated on ice for a further 1 h. The cell pellet is then dispersed by gently rotating the tube on its vertical axis. The cells are then layered onto an equal volume of Ficoll-paque and spun at 400 *g* for 20 min. The interface is then collected and washed three times with RPMI-1640. The remaining cells comprise monocytes and B-lymphocytes.

(iii) B-lymphocytes expressing the desired antibody specificity can be enriched. This procedure can be coupled with the previous one for removing T-cells although this is not strictly necessary. Each antigen will dictate its own selection protocol while some may prove unamenable to enrichment procedures. In the case of the latter, then procedures (i) or (ii) should be adopted. For the Rhesus D antigen, appropriate B-cells can be conveniently selected for by a simple rosetting procedure. The procedure is essentially identical to that described for T-cell depletion except that washed red blood cells from a RhD positive (blood group O) individual are used to form the rosettes and the subsequent incubation at 37°C can be omitted. Also, the cells which pass through the gradient, rather than those which remain at the interface, are the ones which should subsequently be infected with EBV.

### 3.3 **Infection of cells with EBV**

(i) Pellet and re-suspend cells from procedures (i)–(iii) above in virus-containing supernatant to a concentration of $2 \times 10^6$/ml.

(ii) Incubate the cells with the virus in a 37°C waterbath for 1.5 h with mixing every 15 min.

(iii) Wash the cells once and re-suspend in fresh full medium (RPMI-1640, 10% FCS, antibiotics and fungicides) for culture. If B-enriched populations are used then it is advised to include 2-mercaptoethanol in the full medium at a concentration of $5 \times 10^{-5}$ M.

### 3.4 **Propagation of EBV-transformed cells**

Following infection with EBV, cells can be expanded by *in vitro* propagation. Seed infected cells at $2 \times 10^6$/ml of full medium in a suitable sized tissue culture vessel. Flat-bottom vessels are beneficial as these allow easy visual assessment of the progress of the cell line. Remember that if cells have not been T-depleted, include phytohaemagglutinin (e.g. Gibco, UK at 1:100) during the first week of culture. Over the first 4 days it can be observed that some cells form large clumps while others are clearly dying. The cells which emerge represent B-lymphocytes which have been successfully transformed.

After a week or two in culture it is worthwhile checking the supernatant for production of the antibody of interest. Clearly, if a negative result is obtained it is not worth proceeding. Given a positive result then there are two major options open with regard to the strategies subsequently adopted.

**Table 1.** Strategy for production of monoclonal antibodies from bulk cultures.

| | |
|---|---|
| 1. | Allow EBV-transformed cells to grow for 2–3 weeks. |
| 2. | Check antibody production. |
| 3. | Seed two flat-bottom microtitre plates (growth area = 0.32 $cm^2$/well) with $10^4$ cells per well. |
| 4. | Grow for 3–5 days. |
| 5. | Score for antibody production. |
| 6. | At this stage, wells can be sub-cultured onto feeder-layers at numbers ranging from 10 to 1000 cells per well. |
| 7. | Allow another 3–5 days of growth (possibly longer for wells seeded at very low numbers) and score for antibody production. |
| 8. | At this stage wells can be checked for clonality [e.g. by isoelectric focusing (IEF)] or sub-cloned further if warranted. |

### 3.5 Propagation of antibody-producing lines by antigen enrichment

An early observation by workers attempting to produce monoclonal antibodies via EBV transformation was that antibody production in bulk cultures was rapidly lost in favour of cells synthesizing immunoglobulins of irrelevant specificities. To obviate this problem, attempts have been made to enrich for those cells whose immunoglobulins define the correct antigen specificities (10). In theory, this should eventually lead to monospecific lines emerging or, at least, to the propagation of the antibody-producing cells of interest. Several reports suggest that this approach can be successful.

In the case of the RhD antigen, the rosetting procedure described previously can be adopted for selecting the antibody-producing cells of interest. Again, other antigenic systems will warrant their own specific enrichment strategies and these will be considered later. Antibody production can be propagated by this means until it is deemed that monoclonality or, at least, monospecificity has been achieved.

### 3.6 Cloning of antibody-producing lines

As an alternative to the above approach, or even to complement it, propagation and monoclonality of antibody-producing lines can be sought by simple cloning procedures. Some workers clone at the earliest possible stages, that is, directly following infection of cells with EBV (11). Others allow the EBV-infected cells to expand and then clone (12). The advantage of cloning is that it ensures monoclonality of antibody-producing lines from the onset. An overwhelming disadvantage, particularly in the absence of any enrichment procedure, is that the numbers of wells needed to be screened at any one stage may be alarmingly high. Clearly, a compromise is called for. The strategy given in *Table 1* for obtaining Mab production from bulk cultures is offered as an example of the balance which can be struck in attempts to obtain monoclonal populations while limiting the number of wells handled. This outline is best incorporated with antigen enrichment prior to EBV infection where possible. Otherwise, the reader is advised to begin with T-depleted populations as outlined earlier.

### 3.7 Stabilization of antibody-producing clones

Occasionally cloned lines are generated following EBV transformation which are truly immortal. It is the experience of many, however, that a large number of EBV-transformed clones succumb to decay either in their growth or in their ability to secrete antibody (12). Furthermore, even where clones remain stable, the resulting level of antibody production can be disappointingly low. It is for these reasons that increased attention is being focused on the possibility of fusing EBV-transformed cells with already established cell lines in order to generate stable, high-rate antibody-producing clones.

The principles for hybridization are identical to those described previously for rodent monoclonal antibodies (Chapter 2). The candidates for fusion partners were described earlier.

Unlike normal immune cells, EBV-transformed cells display considerable growth potential of their own, so that selection of hybridomas can be problematical. However, as the rationale for hybridization arises from clonal instability or low-rate antibody production, then these properties in themselves can be used to select out non-hybridized antibody-producing cells. If desired, successful hybridization can be verified by karyotypic analysis. Often, it is sufficient simply to observe the morphological changes associated with hybridization of EBV-transformed cells to mouse, human or heteromyelomas. Ideally, fusions should be constructed with partners carrying a resistant marker such as ouabain.

### 3.8 **Harvesting of monoclonal antibodies**

Unlike rodent hybridomas there is no possibility of growing human lines as ascites in syngeneic hosts. The feasibility of propagating human lines in nude mice and rats is being investigated however. At present though, most human Mabs are harvested directly as supernatant from *in vitro* cultures. For diagnostic purposes it is probably sufficient to pass supernatants through 0.2 $\mu$m filters in order to remove the small amount of virus which can be associated with latently-infected lymphoblastoid lines.

Where *in vivo* administration is to be considered, then clearly more critical attempts must be made to neutralize potentially harmful effects of EBV (particularly worrying where immunosuppressed patients are involved). One suggestion is to expose the culture supernatants or antibody fractions to sufficient UV light to inactivate the virus.

## 4. ALTERNATIVE APPROACHES FOR THE PRODUCTION OF HUMAN MONOCLONAL ANTIBODIES

The previous section serves as a guideline for those interested in producing Mabs of human derivation. Alternative approaches exist and the reader should be aware of these. The major alternatives place the emphasis on hybridization rather than EBV transformation. In some studies, freshly isolated peripheral blood cells have been fused directly with permanent cell lines (13). In others, fusion has been performed following stimulation of primary cells with mitogens (14). It appears that the B-blasts may be more amenable to hybridization than the resting cell. Neither of the above approaches takes account of the antigen specificity eventually required following fusion. In attempts to increase the number of precursor cells of correct specificity some workers have adopted a strategy of *in vitro* immunization prior to fusion, apparently with some success (15). Such an approach could be coupled with enrichment procedures to increase the probability of an appropriate fusion even further.

As mentioned earlier, each antigenic system will demand its own enrichment strategy. The possibility of isolating the correct B-cells by rosetting has been discussed for the RhD antigen and such a procedure lends itself readily to other blood group antigens. An alternative approach is to use fluorescein-labelled antigen, as has been described for tetanus toxoid, and then separate out appropriately labelled cells on a fluorescent-activated cell sorter (16). Sufficient ingenuity exists among researchers that if the problem is urgent enough, means will be devised to enrich cells for whatever antigen specificity is desired.

## 5. REFERENCES

1. Gordon,J., Abdul-Ahad,A.K., Hamblin,T.J., Stevenson,F.K. and Stevenson,G.T. (1984) *Br. J. Cancer,* **49**, 547.
2. Steinitz,M., Klein,G., Koskimes,S. and Maekla,O. (1977) *Nature,* **269**, 420.
3. Steinitz,M., Sappela,I., Eichmann,K. and Klein,G. (19797 *Immunology,* **156**, 41.
4. Robinson,J. and Miller,G. (1975) *J. Virol.,* **15**, 1065.
5. Pickering,J.W. and Gelder,F.B. (1982) *J. Immunol.,* **129**, 406.
6. Kozbor,D., Lagarde,A.E. and Roder,J.C. (1982) *Proc. Natl. Acad. Sci. USA,* **79**, 6651.
7. Kozbor,D., Roder,J.C., Chang,T.H., Steplewiski,Z. and Koprowski,H. (1982) *Hybridoma,* **1**, 323.
8. Teng,N.N.H., Kaplan,H.S., Hebert,J.M. *et al.* (1985) *Proc. Natl. Acad. Sci. USA,* **82**, 1790.
9. Moss,D.J., Scott,W. and Pope,J.H. (1977) *Nature,* **268**, 735.
10. Boylston,A.W., Gardner,B., Anderson,R.L. and Hughes-Jones,N.C. (1980) *Scand. J. Immunol.,* **12**, 355.
11. Rosen,A., Persson,K. and Klein,G. (1983) *J. Immunol.,* **130**, 2899.
12. Melamed,M., Gordon,J., Ley,S.C., Edgar,D.H. and Huges-Jones,N.C. (1985) *Eur. J. Immunol.,* **15**, 742.
13. Croce,C.M., Linnenbach,A., Hall,W., Steplewski,Z. and Koprowski,H. (1980) *Nature,* **288**, 488.
14. Chiorazzi,N., Wasserman,R.L. and Kunkel,H.E. (1982) *J. Exp. Med.,* **156**, 930.
15. Kozbor,D. and Roder,J.C. (1984) *Eur. J. Immunol.,* **14**, 23.
16. Kozbor,D. and Roder,J.C. (1981) *J. Immunol.,* **127**, 1275.

CHAPTER 5

# Immunoaffinity chromatography

JOSIANE ARVIEUX and ALAN F.WILLIAMS

## 1. INTRODUCTION

The purification of polyclonal antibodies after binding to insoluble antigens was widely studied in the development of affinity chromatography (1) and antibody purification became routine after the discovery of the cyanogen bromide method for coupling ligands to polysaccharide beads, particularly agarose (2,3).

Despite the ease of antibody purification the reverse case, namely the use of an antibody column to purify antigen, was commonly ineffective. Problems were that to produce a specific antiserum, pure antigen was usually needed and the conventional purification that gave the antigen was likely to be more effective than a scheme involving an antibody column constructed with polyclonal antibodies raised against the antigen. One reason for this is that specific antibodies would only be a minor fraction of the immunoglobulin on a column.

Columns containing polyclonal antibodies raised against partially purified antigen were sometimes useful in purification schemes. Such columns gave enrichment of viral enzymes (4), and human interferon (5), and polyclonal antibody columns were used in the primary purification of Thy-1 and HLA cell surface antigens. Thymocyte Thy-1 was purified using an affinity column containing antibodies raised against semi-pure brain Thy-1 antigen (6) and the intact dimeric form of HLA was purified using specific antibodies against the $\beta_2$-microglobulin chain of the dimer which could itself be purified from urine of patients with kidney disease (7).

The use of polyclonal antibody columns to remove a minor contaminant from a preparation was more straightforward provided that the contaminant itself could be easily purified (8). A specific serum of high antibody content could be raised or antibodies could be purified, and usually column capacity would not be a problem as the material to be removed would be present in fairly small amounts.

The problem of obtaining specific antibody without pure antigen was solved by the monoclonal antibody (Mab) technique (9) which allowed the isolation of antibodies against individual components of a complex mixture (10). Mabs were soon applied in affinity chromatography firstly in the purification of cell surface antigens (11,12) and later to a wide variety of proteins (13,14).

Apart from solving the specificity problem there are two other key differences between Mabs and conventional antibodies.

(i) In a Mab preparation virtually all immunoglobulin is the desired antibody (even if the Mab is from ascitic fluid) compared with a maximum of about 20% and usually much less for conventional antibodies. This means that affinity columns with reasonable capacities can be constructed even though the capacity may be far below the theoretical value (see below).

(ii) Each Mab has a unique set of kinetic and affinity properties and also will usually recognize only one determinant on an antigen.

The individual properties of Mabs can be an advantage because an antibody with desirable properties can be sought by appropriate screening (although a rare Mab with the 'wrong' properties is a considerable irritation!). The absence of binding to more than one determinant may also seem to be important in that elution from the column might be more efficient than if binding occurred to a number of determinants. However in most cases with polyclonal antibodies, a specific antibody molecule would be surrounded on a matrix by inactive immunoglobulin (Ig), and thus binding to more than one determinant would not usually occur.

In practice Mabs of sufficient affinity invariably make useful affinity columns in contrast to polyclonal antibodies which were rarely useful for purification. In our view the key reasons for this are that Mabs can be prepared against impure antigen and that large amounts of the antibody can be easily obtained.

In this chapter we will first give methods for purification of rabbit anti-Ig polyclonal antibodies for use as second antibodies in serology to detect the binding of Mabs, secondly we will discuss Mab affinity chromatography. It is impossible to discuss all applications and we will concentrate on principles and schemes for the purification of cell surface antigens in the hope that these will also cover the main points relevant to other applications. The only methods that we present in detail are those that have been used extensively in our laboratory but key references to other methods will be made wherever possible.

## 2. BUFFERS

### 2.1 DAB

This is Dulbecco's A + B physiological buffer from Oxoid Limited, Basingstoke, UK.

### 2.2 Tris/saline buffers

These buffers contain 25 mM Tris−HCl pH 7.4, 0.02% $NaN_3$ plus NaCl at the concentration given in each case.

### 2.3 Tris/deoxycholate buffers

These contain 10 mM Tris−HCl pH 8.2 (at room temperature, 8.5 at 4°C), 0.02% $NaN_3$ plus Na deoxycholate at the concentration given.

### 2.4 Tris buffer pH 8

This contains 10 mM Tris−HCl pH 8, 0.02% $NaN_3$.

### 2.5 High pH elution buffer

This is made up of 50 mM diethylamine−HCl pH 11.5 and 0.5% Na deoxycholate.

## 3. PURIFICATION OF RABBIT ANTI-IMMUNOGLOBULIN ANTIBODIES

Purified anti-Ig antibodies are the key to serology appropriate for Mabs as illustrated in *Figure 1* (15). Such reagents are now commercially available but can be very expensive particularly if used in rosetting or panning techniques for purification of cells. Pure anti-Ig antibodies must be degraded to $F(ab')_2$ if quantitative assays are required or if Fc receptors are a possible problem (16). It may be that Mabs will replace polyclonal anti-Ig second antibodies but the polyclonal antibodies give higher labelling than any one Mab because more than one molecule binds per molecule of first antibody. In practice suitable mixes of Mabs to replace the rabbit anti-Ig reagents are not yet available. Depending on the application it may be desirable to have rabbit antibodies that are specific for Fab, L chain or Fc of the first antibody and cross-reaction with Ig of the target tissue may be a problem. This applies particularly if mouse Mabs are used on rat tissue or vice versa, since rabbit anti-Ig antibodies give considerable cross-reaction between mouse and rat Ig. Against other species, rabbit anti-rodent Ig usually gives little cross-reaction and the cross-reacting antibodies are easily removed with an appropriate affinity column or blocked by adding IgG or serum to the assay. Below we describe a protocol for the preparation of rabbit anti-mouse IgG antibodies that are devoid of antibodies that cross-react with rat IgG.

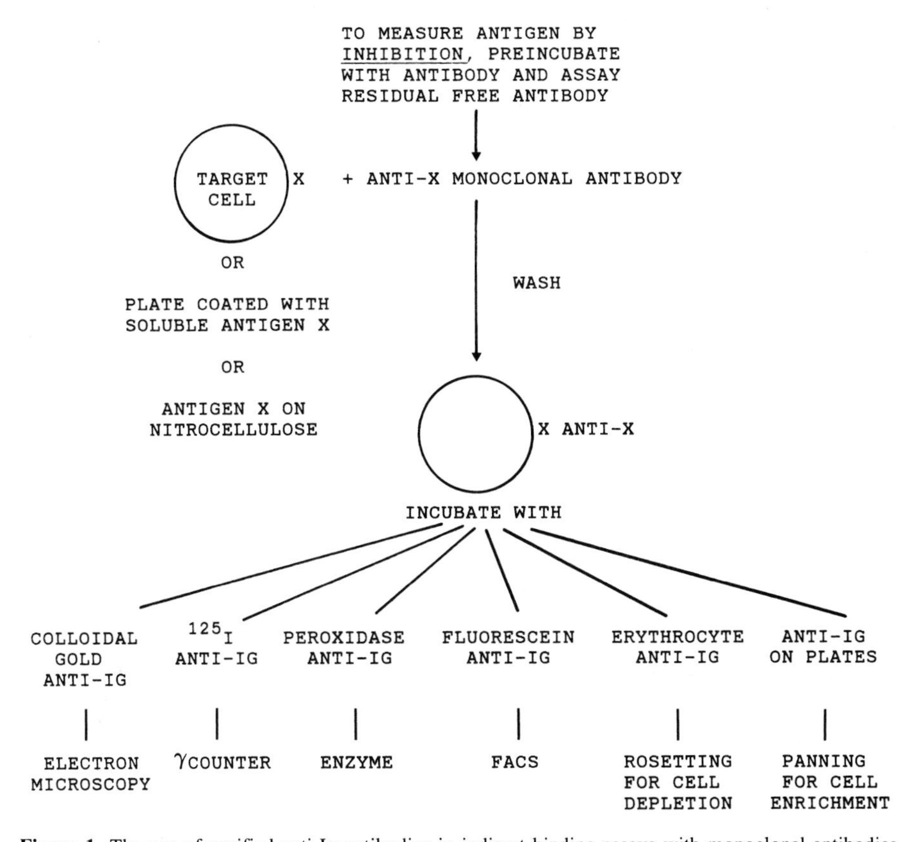


**Figure 1.** The use of purified anti-Ig antibodies in indirect binding assays with monoclonal antibodies.

### 3.1 **Purification of mouse or rat IgG**

See Chapter 2. We use the same protocol for both mouse and rat IgG and make large preparations whenever possible. In brief the protocol is as follows.

(i) Take 100 ml of mouse or rat serum, centrifuge (>10 000 *g*) for 30 min, take the supernatant, add $Na_2SO_4$ to 18% w/v and incubate at 37°C for 30 min.
(ii) Collect the pellet by centrifugation at 5000 *g* for 15 min at 25°C.
(iii) Suspend the pellet to 33 ml in water, add $Na_2SO_4$ to 16% w/v and repeat as above.
(iv) Suspend the pellet in 25 ml of water and dialyse three times against 1 litre of Tris/50 mM saline at 4°C.
(v) Remove the material from dialysis and check that the pH and conductivity is the same as the starting buffer.
(vi) Centrifuge to remove any precipitate and read absorbance at 280 and 260 nm with a 1 cm path length. The ratio 280/260 should be about 1.6, and the protein content in mg/ml is calculated from the absorbance at 280 nm divided by 1.4. From 100 ml mouse serum, 360 mg of protein can be obtained.
(vii) Load the material onto a 50 ml DEAE-cellulose column (Whatman DE-52) in the above buffer and collect the unretarded fraction which should contain about 180 mg of protein. This will be predominantly IgG but can be further purified by gel filtration on a Sephacryl S300 column in Tris/150 mM saline buffer.

The purity of IgG is monitored by polyacrylamide gel electrophoresis in sodium dodecyl sulphate (SDS−PAGE) in reducing and non-reducing conditions. A mouse IgG preparation will contain $IgG_1$, $IgG_{2a}$ and $IgG_{2b}$, while the rat IgG will consist mainly of $IgG_{2a}$, $IgG_{2b}$ and $IgG_{2c}$. In each case these are the important IgG classes for Mab work.

### 3.2 **Immunization**

Use the purest preparations of IgG for immunization.

(i) Inject four rabbits intramuscularly with 0.5−1 mg of mouse IgG per injection in complete Freund's adjuvant (Chapter 2) once in the left and once in the right hind leg at an interval of 1 week.
(ii) Test bleed 1 month after the first injection and assay for antibody by double diffusion in agar gel (Chapter 6) or by indirect binding assays on plastic microtitre plates coated with mouse IgG (Volume II).
(iii) If antibody content is low, immunize again with 0.5−1 mg per rabbit of alum-precipitated IgG (Chapter 2) via intraperitoneal injection. Alternatively, boosts can be given subcutaneously in incomplete Freund's adjuvant or by intravenous injection without adjuvant. Antiserum should preferably contain about 2 mg/ml of anti-IgG antibody (see Volume II for quantitation of antibodies in serum).
(iv) Collect a large volume of serum (e.g. 500 ml) and heat at 56°C for 30 min followed by centrifugation at 40 000 *g* for 30 min prior to affinity chromatography.

### 3.3 **Antibody purification** (16)

All steps are performed at 4°C.

(i) Couple mouse and rat IgG to Sepharose CL-4B beads (see below) at 10 mg IgG/ml and prepare columns containing at least 20 ml of beads (we usually use 40 ml columns). Wash the columns with 150 ml of 1 M propionic acid and then with Tris/500 mM saline buffer.

(ii) Pass the serum through the rat IgG column at about one column volume/h and check for depletion of cross-reacting antibody on emerging fractions. Then pool the fractions and pass through the mouse IgG column. Assay for antibodies on emerging fractions and continue until the column is saturated. Wash both columns with Tris/500 mM saline buffer until absorbance at 280 nm, 1 cm is less than 0.05.

(iii) Elute each column with 1 M propionic acid and read the absorbance of the fractions as they emerge: cloudiness may be observed in fractions with the most antibody. Neutralize immediately with solid Tris. Re-equilibrate the column with Tris/150 mM saline buffer when elution is complete. If $F(ab')_2$ is not wanted then dialyse eluted antibody against Tris/150 mM saline or another buffer of choice. From a column that was saturated with antibody the weight of pure IgG eluted is usually roughly equal to the amount of IgG on the column.

(iv) To prepare $F(ab')_2$, dialyse the eluted antibody against 0.1 M Na acetate buffer pH 4.4 and add pepsin to 2% w/w (antibody should be >10 mg/ml; if not, concentrate by ultrafiltration). Incubate at 37°C for 6–8 h and then add a further 1% w/w pepsin and incubate for a further 12 h (alternatively, to be sure of the outcome a trial incubation could be carried out for various times with the results assayed by SDS–PAGE with unreduced and reduced samples). At the end of the incubation bring to pH 7 with Tris and separate $F(ab')_2$ from other fragments by gel filtration on Pharmacia Sephacryl S200 in Tris/150 mM saline buffer. Monitor the gel filtration by absorbance at 280 nm and SDS–PAGE. The yield of $F(ab')_2$ is usually 50% by weight of the IgG digested, which is 75% of theoretical.

(v) Concentrate to the desired protein concentration by ultrafiltration and store at −20°C.

## 4. MONOCLONAL ANTIBODY AFFINITY CHROMATOGRAPHY

### 4.1 Antibodies suitable for affinity chromatography

For affinity chromatography the antibody on the matrix is required to bind free antigen in solution; thus usually the key parameter is the affinity of the monovalent interaction (17). Kinetics of reaction are not usually a problem since the Mab is coupled to the beads at high concentration but in exceptional cases a column might not effectively bind antigen at the usual flow-rates of about one column volume/h. Assays for Mabs often allow multivalent binding (see *Figure 1*) leading to detection of both low and high affinity antibodies. A second screen is thus needed to find Mabs suitable for affinity chromatography and this must be based on monovalent interactions. The ability to immunoprecipitate antigen provides one such test, and another is the demonstration that soluble antigen can function to inhibit antibody binding to multivalent antigen at a cell surface or on a plate (17). The possible events occurring in the inhibition assay are summarized in *Figure 2*; it can be seen that blocking will only occur if the monovalent rate of dissociation is such that the time for 50% dissociation is roughly equal to, or

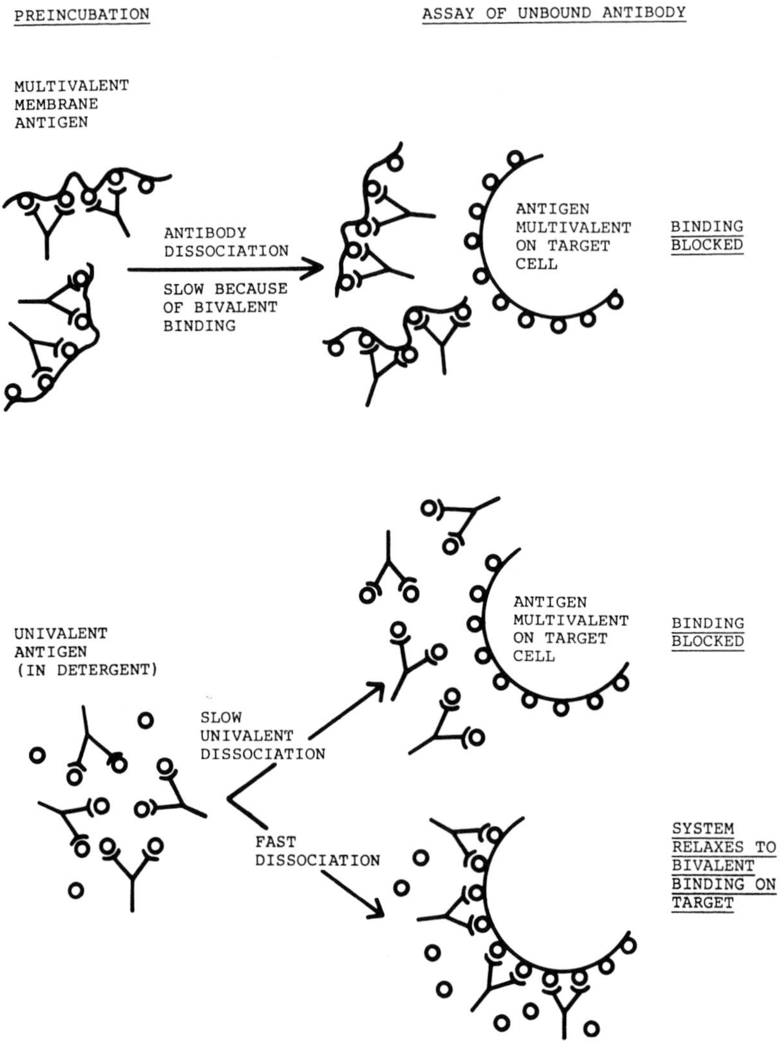

**Figure 2.** Effect of dissociation rate on the inhibition of antibody binding with univalent or multivalent antigen.

longer than, the time for the binding assay. Antibodies with $t_{1/2}$ greater than 10 min will usually have affinities for protein antigens of greater than $10^8$ $M^{-1}$ and this is sufficient for affinity chromatography (18).

To ensure that high affinity Mabs can be selected it is best to hyperimmunize animals such that a strong secondary response dominated by high affinity IgG antibodies is elicited. To our knowledge IgM antibodies have not been useful in affinity chromatography.

It could be supposed that very high affinity antibodies should be avoided because of the possibility that antigen will not be readily released in the elution step. However it is not obvious that the affinity in dissociating conditions correlates with that seen in physiological (i.e. binding) conditions, and the main criterion with regard to affinity is that it should be high enough to allow effective retention of antigen during application of the antigen-containing extract to the column, and throughout the washing steps.

With Mabs against cell surface antigens it is important to consider whether a protein or carbohydrate determinant is being recognized. A protein determinant is usually unique to one molecule and even if cross-reaction with an irrelevant molecule happened to occur, the affinity to this would probably be too low to afford any confusion. Carbohydrate structures, however, are usually shared between a number of glycoproteins and/or glycolipids and thus unique structures would not be purified. This is not usually a problem with Mabs against mammalian glycoproteins since antibodies to protein determinants predominate probably because most carbohydrate structures are common across the mammalian species and thus are not seen as foreign structures in the immunized animal. However some carbohydrates can be very immunogenic in nature; for example most Mabs raised against a slime mould glycoprotein recognized a carbohydrate determinant common to a multiplicity of different membrane molecules (19).

### 4.2 Isolation of monoclonal antibody IgG

We isolate Mab IgG from ascitic fluid. This should contain more than 2 mg/ml of the Mab. The steps used are the same as for mouse serum (Section 3.1) except that we use 18% w/w $Na_2SO_4$ at both salt precipitation steps and do not carry out the gel filtration step. The Mab activity must be assayed at all stages since variations from the norm can occur at both the $Na_2SO_4$ and the DEAE-cellulose steps. Sometimes the level of $Na_2SO_4$ is raised to 20% w/w and some Mabs may bind to DEAE in Tris/50 mM saline. If the latter occurs, elution can usually be obtained with Tris/75 mM NaCl buffer. For many applications semi-pure Mab IgG, resulting from $Na_2SO_4$ precipitate alone, may be used but if column capacity is to be maximal the purer IgG resulting from the DEAE-cellulose step should be used. Contamination of Mab IgG with normal IgG is usually less than 10% provided that the hybridoma being used gives 100% positive wells on re-cloning. Prior to coupling we dialyse IgG at about 10 mg/ml against 0.05 M disodium tetraborate:HCl buffer pH 8.

### 4.3 Coupling of monoclonal antibodies to agarose beads

**NOTE: CYANOGEN BROMIDE IS TOXIC AND VOLATILE**

We always use the cyanogen bromide (CNBr) coupling method with Pharmacia agarose beads. Other coupling procedures are discussed elsewhere (20). Beads pre-activated with CNBr can be purchased and used according to the manufacturer's instructions but coupling according to Porath (21) is simple, reproducible and cheap. We have used Pharmacia Sepharose 2B, 4B and CL-4B but currently use only the cross-linked Sepharose CL beads. It has been recommended that Mabs should be attached to Protein A-Sepharose 4B and then cross-linked with glutaraldehyde (22). It is not obvious that a spacer group should be needed for antibodies since attachment via one Fab or Fc should leave at least one Fab free for binding. Calculations from the results in reference 22 did not suggest a greater capacity than we find with direct coupling but the Protein A method may have advantages in some cases.

Antibody is coupled at 5 – 10 mg/ml of swollen beads with the minimum of CNBr needed for 100% coupling. For consistent results, good quality CNBr must be used and we aliquot a new 25 g batch in the fume hood into about 16 pre-weighed glass bottles (5 ml) with screw caps and determine the weight of CNBr in each by weighing the sealed bottles. These are then put in a container with desiccant which is prominently labelled, sealed with a lid and tape and stored at −20°C. For use the whole container is warmed in the fume hood and an aliquot removed thus ensuring that CNBr is always in a sealed container unless being used in the fume hood.

Steps in coupling of antibody to beads are as follows with steps (ii)−(vi) done in the fume hood.

(i) Wash Sepharose in water and measure the packed volume by centrifugation in a calibrated centrifuge tube (500 *g* for 5 min).

(ii) Place 1 vol of beads, 0.5 vol of water, in a beaker (capacity 5 vol) with a stirrer set up in an ice bath on a magnetic stirrer. Add 1 vol of ice-cold 4 M $K_2HPO_4$/KOH pH 11.5 when CNBr is ready to be added.

(iii) Take CNBr at a level of 30 mg/ml of packed beads for Sepharose 4B, or 40 mg/ml for Sepharose CL-4B and dissolve in dimethylformamide at 300 mg/ml. Add the solution to the beads as in (ii) over 1 min and stir for a further 9 min.

(iv) Pour the slurry quickly into a sintered funnel and suck-off the supernatant under vacuum via a water pump. Wash with about 5 vol of ice-cold water and repeat this 4−5 times allowing the beads to be sucked dry after each addition.

(v) Transfer the beads to the IgG solution in 0.05 M disodium tetraborate:HCl buffer pH 8 in a screw-capped plastic tube or centrifuge pot (the beads should take up most of the fluid), and rotate slowly at 4°C for 2 h or mix every 10 min for 2 h. Leave to stand overnight at 4°C.

(vi) Place all used glassware into commercial bleach in the fume hood as the preparation goes on to neutralize CNBr, and then wash away residues with copious water.

(vii) Next morning centrifuge the beads at 500 *g* for 5 min and determine the absorbance of the supernatant at 280 nm to measure the amount of uncoupled IgG. We routinely obtain more than 90% coupling efficiency. Wash the conjugated beads on a funnel with Tris/150 mM saline and then leave to stand for 10 min with 50 mM ethanolamine−HCl pH 8 to block any remaining active groups. Store the beads in the above Tris buffer at 4°C.

(viii) Before using the column, pre-elute with the buffer that is later to be used for antigen dissociation.

### 4.4 Affinity columns and flow-rates

We use simple affinity columns run under gravity and constructed in the barrels of glass syringes, as shown in *Figure 3*. With this construction, columns of volume from 1 ml to 50 ml are possible and this covers most applications. We usually use columns of 5 ml or greater since the yield of a column has not been observed to fall if the column is not saturated with antigen. A guard column containing 10−20 ml of bovine $\gamma$-globulin−Sepharose CL-4B is set up above the specific column. The necessity for this is not proven but the aim is to trap any particles or materials that would bind other

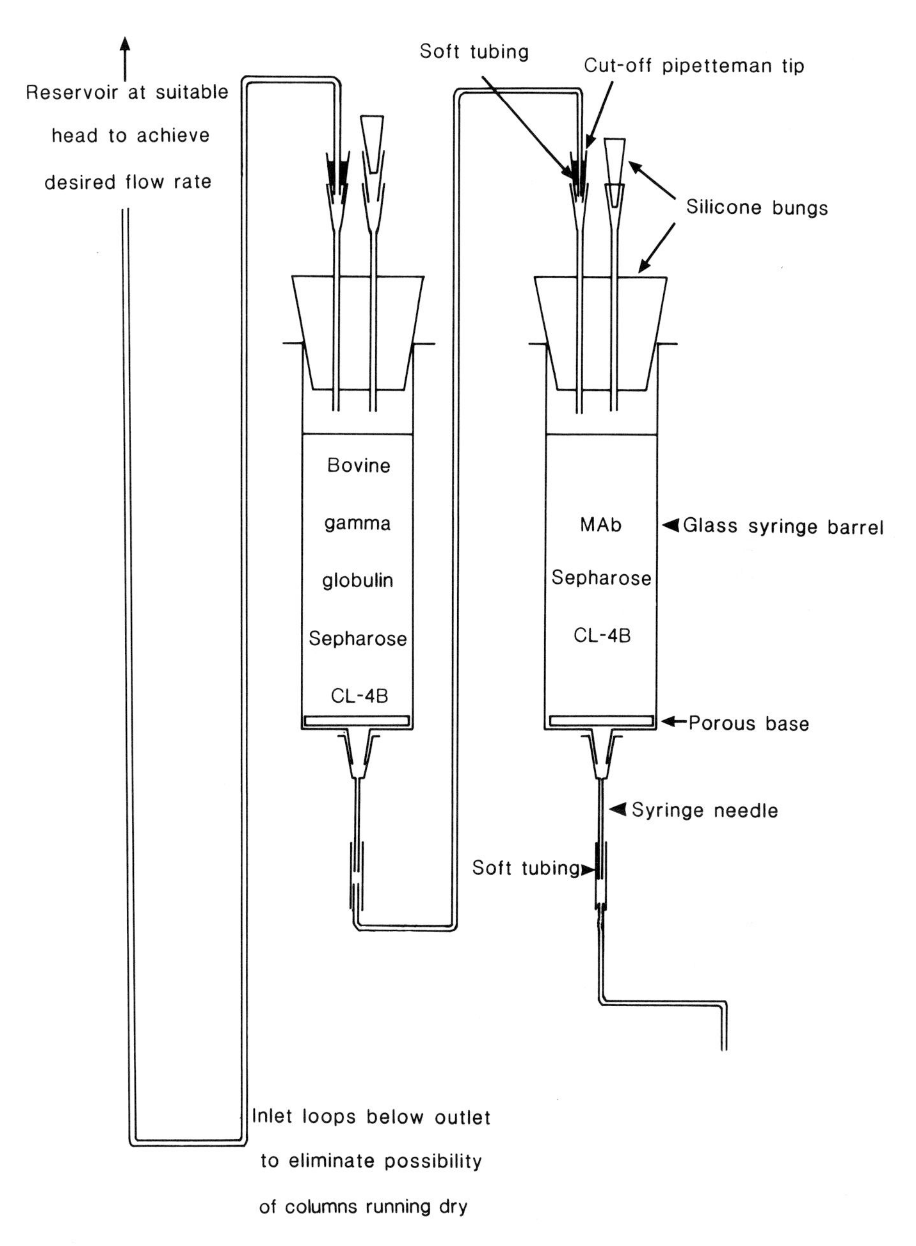


**Figure 3.** Arrangement of affinity columns.

than by the antibody combining site. Flow-rates of 1–2 column volumes per hour are usually used and it is best not to pump extracts through columns, since sometimes resistance of the columns increases and this could lead to a blow-out and loss of material. Columns can be washed at higher flow-rates.

### 4.5 **Batch method**

As an alternative to passing extract through the column, a large volume of extract can be incubated with the Mab–Sepharose CL-4B, with rotation overnight at 4°C. The beads are then filtered on a funnel and packed in a column for washing and elution. The main advantage of this procedure is that a large volume of extract may be depleted of antigen more quickly than by passage through the column.

### 4.6 **Washing**

After all the extract has passed, the column is washed with buffer until the absorbance at 280 nm equals the buffer absorbance. The only key criterion for a washing buffer is that it should not elute the antigen. However, conditions likely to displace absorbed protein can be used. For example in purifying cell surface antigens we would first wash with Tris/0.5% deoxycholate buffer and then with the same buffer plus 0.15 M NaCl (this is made up immediately before use since deoxycholate forms a gel in the presence of salt if left overnight at 4°C). If contaminants are a problem, buffers with greater dissociating properties can be tried in the wash step.

### 4.7 **Elution of antigen**

Partial denaturation of protein structure is required to release antibody from the antibody combining site but full denaturation, as would be achieved with 8 M guanidine–HCl or by boiling in SDS, is not usually required. This is fortunate since it is desirable that both antigen and antibody should regain a native conformation after the elution steps. Dissociating conditions have most often been evaluated in terms of eluting antibody from insoluble antigen. Commonly used conditions are given below.

(i) *High pH.* This seems the most generally effective method. 50 mM diethylamine–HCl, pH 11.5 is the most commonly used buffer and is compatible with all detergents including bile salts (6,23,24).

(ii) *Low pH.* This was very commonly used in antibody purification (3–5,23). Buffers are 0.1 M glycine–HCl pH 2.2, 0.1 M citrate pH 2.2, 0.5 M propionic acid pH 3. Propionic acid may have the advantage of interfering with hydrophobic interactions as well as having effects via the change in pH. Low pH is compatible with non-ionic detergents but not bile salts.

(iii) *Chaotropic agents.* Anions are effective in breaking protein interactions in the order $SCN^- > ClO_4^- > I^- > Br^- > Cl^-$ and cations in the order $Mg^{2+} > K^+ > Na^+$. 2.5 M KI or 2.0 M KSCN gave effective elution of anti-Rh antibody from erythrocytes (25).

(iv) *Heat.* We usually elute at 4°C but antibody affinity usually falls if the temperature is increased, and elution by temperatures above 37°C may be worth investigating.

(v) *Hypotonic solutions.* Elution with 2 mM Tris–HCl pH 8 has been used for eluting a kidney brush border enzyme (26).

(vi) *Agents that lower surface tension.* High concentrations of ethylene glycol or dimethyl sulphoxide can dissociate antibody complexes. With 60% ethylene glycol, albumin: anti-albumin complexes can be dissociated at pH 3.5 or 9.5 (27).

(vii) *Individual cases.* Sometimes a unique property of the antigen:antibody reaction will allow elution under very mild conditions. With one Mab to an H-2 antigen, binding to a column occurred with non-ionic detergent but elution was possible with deoxycholate even though this does not usually bind to non-hydrophobic parts of a protein (28). Antibodies to clotting factor IX can recognize a determinant that is dependent for conformation on $Ca^{2+}$ ions and with these antibodies elution was possible with a buffer containing EDTA (29).

In practice we usually elute cell surface antigens in 50 mM diethylamine–HCl pH 11.5/0.5% deoxycholate buffer and immediately neutralize with glycine. Absorbance at 280 nm is read and antigenic activity is measured. Usually a yield of about 50% antigen applied to the column is obtained and eluted fractions are concentrated by ultrafiltration and deoxycholate removed by dialysis if required. In one case involving the purification of a T-lymphocyte antigen named MRC OX-34, no 280 nm absorbing material or antigenic activity was eluted from the column even though all antigenic activity had been removed from the detergent extract that was passed through the column. Thus elution with 0.5 M propionic acid with no detergent was tried and with this the antigen was eluted with good yield of antigenic activity (A.F.Williams, unpublished results). Also with the rat CD4 (W3/25) antigen, poor results were obtained with the high pH buffer and in this case 0.1 M glycine–HCl pH 2.5 buffer was used (30). Protein of the correct apparent molecular weight was recovered from the column but antigenic activity was partially lost. Thus in this case the criteria for purification depended on the fact that the antigenic activity has been removed by the column and a protein of the known apparent molecular weight recovered.

In our experience the loss of antigen activity on elution is exceptional and there are also many examples where proteins have retained their enzymatic or binding activities after elution with high or low pH buffers (13,26,31,32).

### 4.8 **Column capacities and re-use**

In our experience maximum column capacity is usually not more than 20% of theoretical with antibody coupled at 10 mg/ml (14). At this level a 10 ml column would bind 13.3 mg (270 nm) of a 50 000 molecular weight antigen. However this capacity is often not achieved and capacities seem to be particularly bad for large molecules and in some cases may only be a few per cent of theoretical. Presumably much of the Mab is not accessible if an antigen with a very large Stoke's radius is involved. As with most aspects of affinity chromatography, systematic studies on capacity have yet to be carried out.

Columns can generally be re-used and are stored in Tris/150 mM saline at 4°C.

### 4.9 **Leakage of antibody**

In some cases we have observed leakage of Mab from the column when antigen is eluted. On the occasions where this occurred the Mab was separated from antigen in a subsequent gel filtration step, or sometimes by passage through an anti-Ig affinity column. In recent purifications, using Mab coupled to Sepharose CL-4B, we have not observed leakage of Mab at all, and it may be that this is not seen with the cross-linked beads. In our experience leakage of Mab has not been a serious problem.

## 5. PURIFICATION OF PLASMA MEMBRANE ANTIGENS

In this section we discuss the purification of cell surface antigens as shown in *Figure 4* and illustrate points with data from the purification of the MRC OX-45 antigen. This is an antigen of leukocytes and endothelium (33), and is of interest here because the purification illustrates results from two tissues, namely spleen and brain. Also the very high purification factors that are possible with Mab affinity chromatography are illustrated by the purification of OX-45 from rat brain, where the antigen is mostly confined to the endothelium of capillaries.

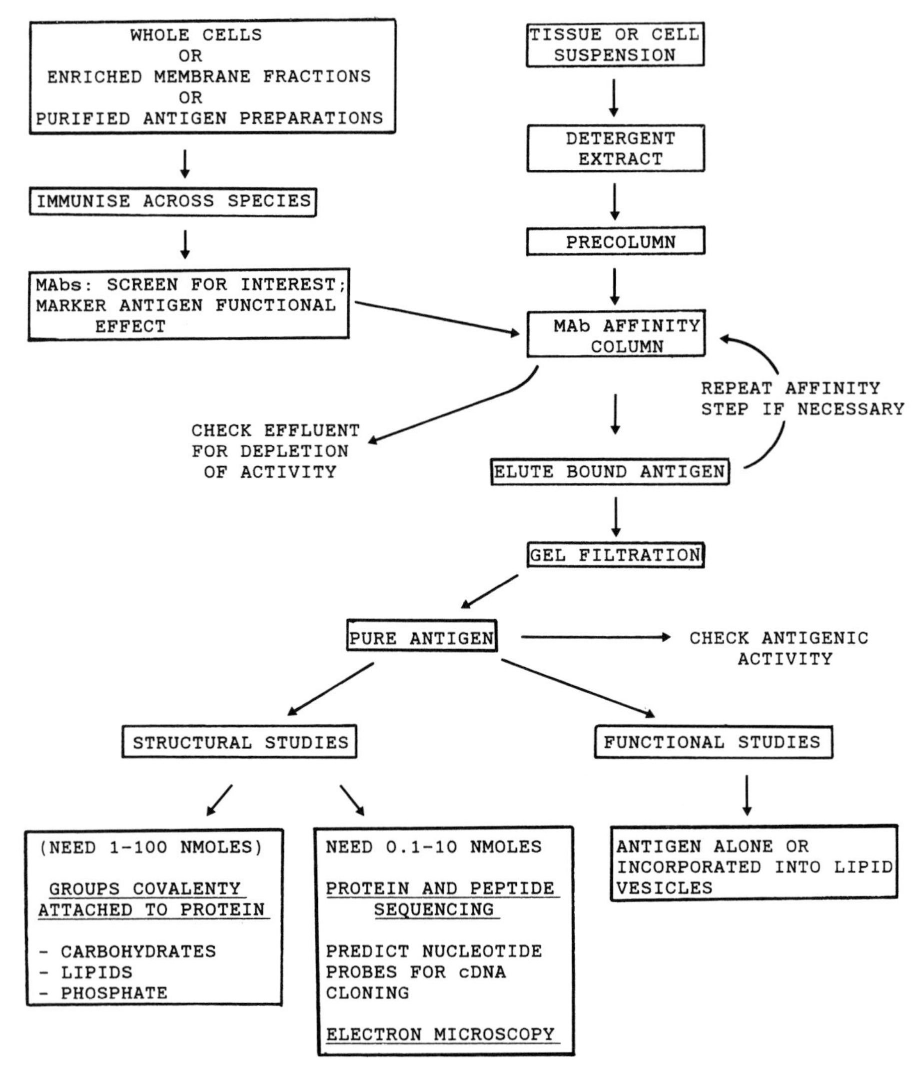


**Figure 4.** Steps in the purification of cell surface antigens by Mab affinity chromatography.

### 5.1 Assays for antigen activity

It is crucial that antigen in extracts is assayed so that it is known whether the activity is being depleted by the column. Also recovery of antigenic activity at a sensible level of purification is a strong assurance that the correct material has been obtained rather than a major contaminant. If activity is lost on elution, careful controls are needed to be sure that the major molecule in the eluted fraction is in fact the antigen.

We use inhibition of indirect binding assays on glutaraldehyde-fixed target cells to follow antigen from starting cells to purified material (15). However an alternative assay is to adsorb material to nitrocellulose at each stage and carry out an indirect binding assay on the adsorbed 'target' material (34).

#### 5.1.1 *Fixation of cells*

(See also ref. 35 for assays with cells fixed to plates.)

(i) Wash the appropriate target cells (lymphocytes, erythrocytes, leukaemic cells, lymphoblastoid cell lines, etc.) once in DAB, resuspend at $10^8$/ml and add an equal volume of 0.25% glutaraldehyde in DAB. Note that tissue homogenates can also be used (36).

(ii) After incubation for 5 min at 23°C, stop the reaction by the addition of 1/10 volume of 10% bovine serum albumin (BSA) and wash the cells once with 0.5% BSA/DAB.

(iii) Resuspend the cells in 10% BSA/DAB, pass them through a syringe needle to avoid clumping and adjust to $5 \times 10^8$/ml for erythrocytes and $10^8$/ml for other cell types in 10% BSA/DAB. Cells can be stored at −30°C for several years without loss of antigenicity.

#### 5.1.2 *Indirect trace binding assay*

All steps are performed at 4°C.

(i) Incubate $2 \times 10^6$ fixed target cells (20 μl) plus $5 \times 10^6$ fixed sheep erythrocytes (10 μl, added to enhance clarity of pellets during the washing procedures) with 25 μl of serially diluted antibody (in DAB/0.5% BSA) in U-bottom microtitre plates with shaking for 1 h.

(ii) Wash the cells three times by centrifugation (500 *g* for 5 min) from 0.1% BSA/DAB (200 μl/well) and then resuspend in 25 μl of the same medium containing $^{125}$I-labelled rabbit $F(ab')_2$ anti-mouse IgG (16) ($2 \times 10^5$ c.p.m./well; specific activity $\sim 4 \times 10^7$ c.p.m./μg).

(iii) After shaking for 1 h, wash the cells three times as above and assay for bound radioactivity. Background values are 500−2000 c.p.m. depending on the target cells used.

(iv) Alternatively carry out the assays in plastic tubes; the procedures are described in detail in ref. 17.

*Figure 5* shows a titration for MRC OX-45 Mab binding to rat thymocyte target cells. A dilution of 1/150 of hybridoma supernatant was chosen for inhibition assays.

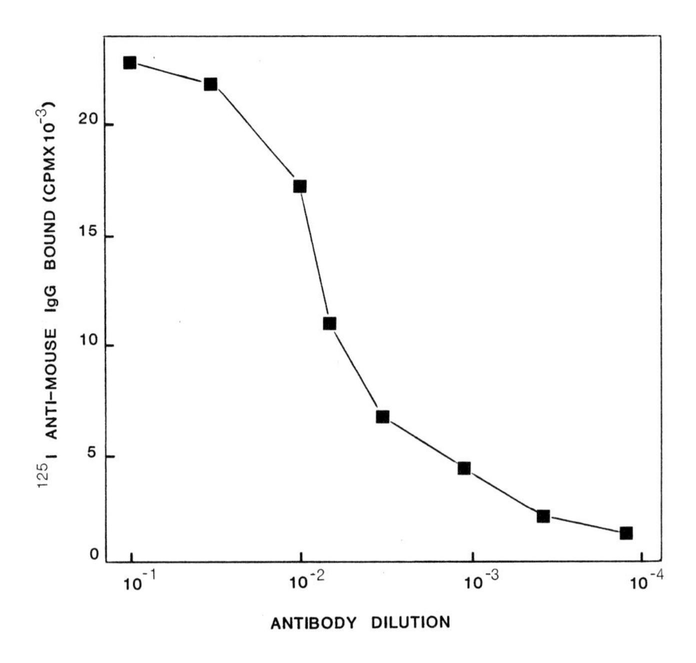


**Figure 5.** Titration of MRC OX-45 antibody in the indirect trace binding assay. Tissue culture supernatant containing antibody was diluted in DAB buffer, 0.02% $NaN_3$, 0.5% BSA and 30 $\mu$l duplicate aliquots were added to the same volume of target cells ($2 \times 10^6$ rat thymocytes + $5 \times 10^6$ sheep erythrocytes) in a microtitre plate. Incubation with $^{125}$I-labelled F(ab')$_2$ rabbit anti-mouse IgG and washings were as described in Section 5.1.2.

### 5.1.3 *Inhibition assays*

Antigen in various tissues can be assayed by absorption analysis where antibody is pre-incubated with tissue homogenate, centrifuged and then the supernatant assayed for residual Mab (*Figures 1* and *2*).

(i) Titrate Mab as above and then use at a suitable dilution in Tris/150 mM saline/2% BSA.
(ii) Make three-fold dilutions of antigen fractions into the above buffer, with 2% BSA if detergents are involved, or 0.5% BSA in the absence of detergent (BSA in the assays, including that from the target cells, adsorbs detergents).
(iii) Incubate equal volumes (e.g. 100 $\mu$l) of diluted Mab plus antigen at 4°C for 2 h (longer or shorter times might be used depending on requirements and assay sensitivity).
(iv) Centrifuge the mixture and take 60 $\mu$l duplicate aliquots for indirect binding assay.

Results for OX-45 Mab absorbed with tissues are given in *Figure 6A* which shows that the antigenic activity per mg of protein for spleen tissue is roughly twice that for thymus and 20 times that for brain. *Figure 6B* shows the absorption of antibody with unsolubilized spleen membrane and with deoxycholate extract prepared from this. The inhibition curve has been displaced somewhat in detergent but the assay is suitable for following depletion and recovery of antigen in the affinity purification procedures.

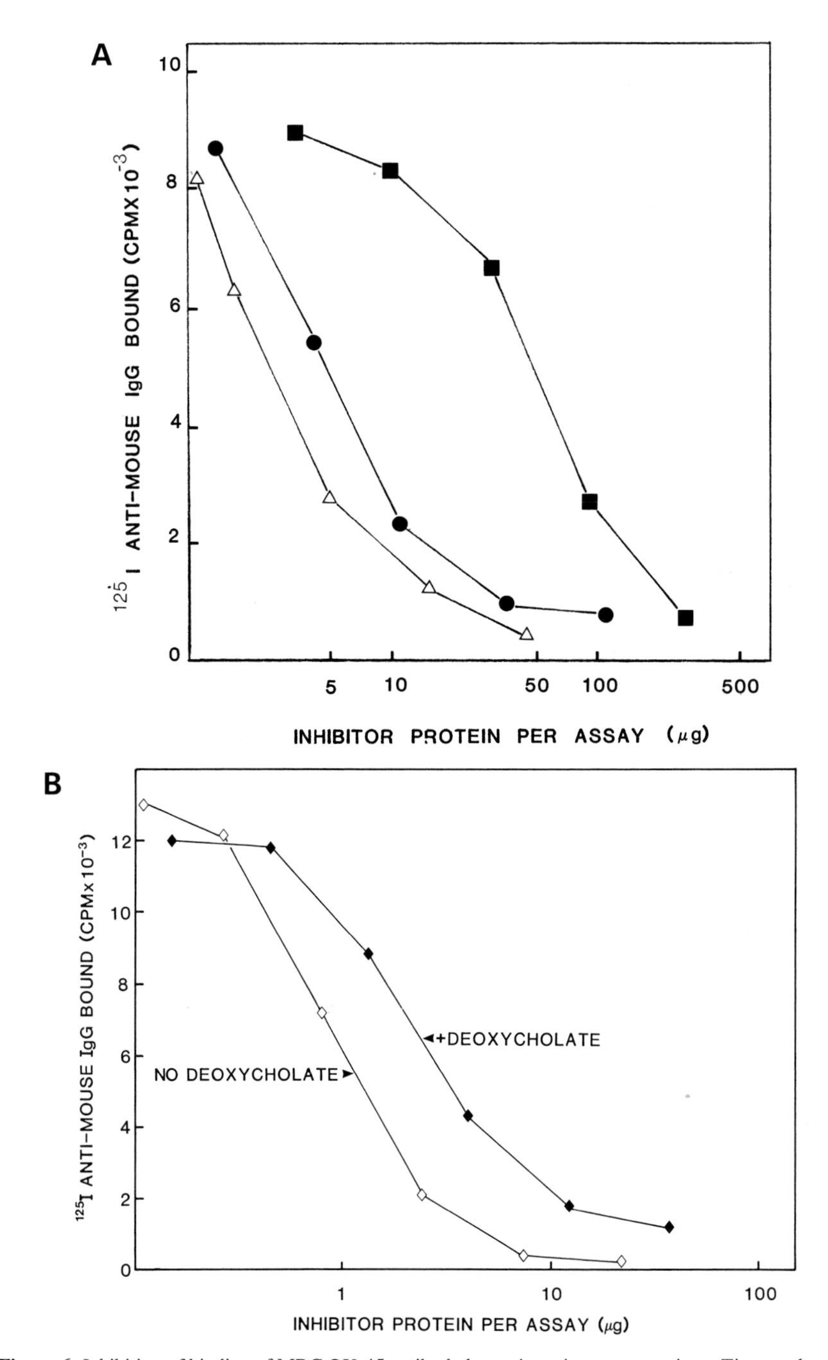


**Figure 6.** Inhibition of binding of MRC OX-45 antibody by various tissue preparations. Tissue culture supernatant containing OX-45 antibody was diluted 1/150 and incubated with an equal volume of materials as described below. After centrifugation the residual antibody was assayed in the binding assay. The details are given in Section 5.1.3. **(A)** Antibody absorbed with (●) thymocytes, (△) homogenate from spleen, (■) homogenate from brain. **(B)** Absorption with (◇) membrane prepared by the Tween-40 method (Section 5.2.4) from rat spleen, (◆) deoxycholate extract from the same membrane (Section 5.2.5).

## 5.2 Preparation of cell extracts

### 5.2.1 *Supply of tissue*

The first consideration is that there is enough tissue for an effective purification and *Figure 4* gives some rough values for the amount of material needed per preparation for different purposes. For an antigen of molecular weight 50 000 which is expressed at $5 \times 10^4$ sites per cell, $6 \times 10^{11}$ cells would be needed to provide 0.5 mg (10 nm) of glycoprotein assuming a 10% yield. This cell number can be obtained from thymuses of 200 rats and larger amounts of non-lymphoid tissues are obtained.

### 5.2.2 *Properties of detergents*

Properties of some detergents used in solubilizing antigens are given in *Table 1*. Generally a useful detergent replaces lipids bound to hydrophobic domains without denaturing the rest of the protein (37). The detergents should be added at a weight to protein ratio of about 10:1.

Aspects relevant to the use of the various categories of detergents are given below.

(i) *Strongly ionic detergents* (*e.g. SDS*). This is not generally useful in affinity chromatography as it binds to all parts of a protein to produce extensive denaturation. However some antigenic determinants are not conformation-dependent and in some cases it is useful to denature in SDS and then, prior to the affinity step, add deoxycholate or non-ionic detergent to form mixed micelles and thus prevent any detrimental effect of SDS on the column. This strategy was used to separate the two chains of the rat CD8 antigen. The disulphide-linked dimer was first purified and then the chains dissociated by reduction in SDS after which excess deoxycholate was added and the affinity chromatography step repeated to separate the chains (38).

(ii) *Weakly ionic detergents.* Deoxycholate usually seems to give the most complete solubilization of membrane molecules and its relatively high critical micellar concentration allows the detergent to be removed by dialysis. Also the small micelle size means that the behaviour of solubilized proteins on gel filtration is not dominated by the bound detergents. We commonly use a gel filtration step after the Mab column to remove minor contaminants. The disadvantage of deoxycholate is that it forms a gel at pH values less than 8, and with the addition of salts. Cholate is much less troublesome but also less effective for solubilization. However it may be an advantage to solubilize in deoxycholate and then change to cholate when antigen has bound to the column. With lymphoid tissue, nuclei must be removed prior to solubilizing with bile salts otherwise the DNA is released to give a highly viscous solution. However with brain an extract which is not unduly viscous can be directly prepared with deoxycholate. This is presumably because there are fewer nuclei per weight of tissue in the brain.

To prepare deoxycholate for use, first purchase a batch with a low absorbance at 280 nm (<0.025 for a 0.05% solution). This can be used directly for most steps but material for the final step in antigen purification should contain no material that is not dialysable and to be sure of this a 5 litre solution of 0.5% deoxycholate can be prepared by dialysis from 250 ml of 10% deoxycholate inside a dialysis bag. We always solubilize deoxycholate in 10 mM Tris–HCl pH 8.2 (room temperature) plus 0.02% $NaN_3$.

**Table 1.** Properties of detergents.

| *Detergent type* | *Micelle size* ($M_r$) | *Critical micellar concentration (mM)* | *Binds to protein* | | *Interference with antigen:antibody* | *Potency for solubilization* | *Optically clear at 280 nm* |
|---|---|---|---|---|---|---|---|
| | | | *Most sequences* | *Hydrophobic* | | | |
| Strongly ionic | | | | | | | |
| e.g. SDS | 18 000–36 000[a] | 0.5–8[a] | + | + | + | ++ | Yes |
| Weakly ionic | | | | | | | |
| e.g. sodium deoxycholate | 17 00 upwards[a] | 4–6 | –[b] | + | – | ++ | Yes |
| sodium cholate | 900–1800 | 13–15 | – | + | – | + | Yes |
| Non-ionic | | | | | | | |
| e.g. Triton X-100 | 90 000 | 0.24 | – | + | – | + | No |
| Brij-96 | Large | <0.04 | – | + | – | + | Yes |

Data from ref. 37; values at room temperature.
[a]Values affected by buffer constituents.
[b]Some interaction with non-hydrophobic sequences may occur.

(iii) *Non-ionic detergents.* These can be used to solubilize directly membrane molecules from whole cells. Disadvantages are that not all membrane molecules are solubilized, that the large micelle size diminishes the usefulness of gel filtration and that the detergents cannot be removed by dialysis.

### 5.2.3 *Proteolytic inhibitors*

We always add proteolytic inhibitors when cells or membranes are solubilized in detergent and sometimes in the course of membrane preparation. To inhibit sulphydryl-type proteases we add iodometamide to give a final concentration of 2.5 mM. This also blocks free sulphydryl groups and prevents disulphide interchange between molecules. Serine proteases appear to be the main source of lymphocyte proteolytic activity (39) and to inhibit this we use phenylmethylsulphonyl fluoride (PMSF) (in the past we have used diisopropyl fluoride but this is more difficult to handle and PMSF seems as effective). PMSF is toxic and is stored in a small desiccator in the fume hood and always handled in a fume hood. For use, a small glass bottle plus cap is weighed and PMSF added in the fume hood followed by re-weighing of the sealed vial. The PMSF is then made up to 100 mM in isopropanol. In the primary addition PMSF is added to 0.5 mM and further aliquots are added at strategic times in the purification to 0.1 mM. This is because PMSF is quickly hydrolysed in water and proenzymes may be activated at various stages in a preparation. We do not retain the PMSF in isopropanol but place the excess in water in the fume hood and then discard it down the sink with a large volume of water.

### 5.2.4 *Preparation of crude membrane*

We believe that the Tween-40 method (40) is the simplest procedure and is carried out as follows with all steps at 0–4°C.

(i) For single cells (e.g. thymocytes): tease into Tris/140 mM saline pH 7.4 buffer and filter through muslin. Cells should finally be at $10^9$/ml and are not washed so that fragments from broken cells are retained. 250 ml is a convenient volume. For solid tissues (e.g. spleen): take 125 g of tissue (or whatever is suitable) and homogenize in 250 ml of Tris/140 mM saline using a Waring blender.

(ii) Prepare 5% Tween-40 in Tris/140 mM saline pH 7.4 buffer (warm to dissolve, then cool) and add an equal volume of this to either of the above cell preparations. Proteolytic inhibitors can be added at this stage, e.g. 2.5 mM iodoacetamide, 0.5 mM PMSF.

(iii) Stir for 60 min and then homogenize with six strokes of a motor-driven Potter–Elvehjem homogenizer, volume 100 ml. Centrifuge at 1500 *g* for 10 min and then take the supernatant, which should contain about 60% of initial antigenic activity. Resuspend the pellet in 2.5% Tween-40 in the above buffer [volume 20% of total volume in (ii)]; homogenize and centrifuge again as above. The supernatant should contain a further 20% of the initial antigenic activity.

(iv) Take pooled supernatant and either centrifuge to pellet a crude membrane fraction or layer over 32% sucrose in Tris pH 8 buffer and centrifuge to give a partially purified membrane. The sucrose can be placed in polycarbonate tubes (20 ml per tube of 32% sucrose) for a Beckman fixed-angle Type 35 rotor or in polyallomer tubes for a SW27 swing-out rotor (10 ml of 32% sucrose). The rotors are centrifuged for 60 min at 30 000 r.p.m. (Type 35) or 25 000 r.p.m. (SW27). If crude membrane has been prepared this is re-suspended by homogenization in Tris buffer, pH 8, and stored frozen. With the sucrose step the membrane floating at the sucrose−buffer interphase is taken with the minimum amount of supernatant, diluted with Tris buffer, pH 8, and then pelleted and suspended as for crude membrane. The recovery of antigenic activity as a percentage of that in the supernatant is about 90% for crude membrane. With the sucrose step about 60% of the activity in the supernatant is recovered. The membrane fractions can be solubilized immediately or stored frozen at −40°C.

5.2.5 *Solubilization of membrane in deoxycholate*

(i) Take membrane at 3−4 mg/ml protein in Tris buffer pH 8 and add an equal volume of Tris/4% deoxycholate buffer plus 5 mM iodoacetamide, 0.5 mM PMSF. It is important that the deoxycholate:protein ratio (w/w) should be at least 10:1. If necessary, the final deoxycholate concentration can be greater than 2% and we have used up to 5% without affecting the affinity chromatography step.
(ii) Homogenize the extract and stir for 60 min at 4°C; remove insoluble material by centrifugation at 70 000 *g* for 60 min.
(iii) Apply the supernatant to the antibody affinity column or store frozen at −40°C. After thawing, add proteolytic inhibitors (iodoacetamide 1 mM, PMSF 0.2 mM) and stir for 60 min and then centrifuge as above. Some insoluble material will always be present after thawing but we have not detected losses of antigenic activity on freezing and thawing.

5.2.6 *Direct solubilization from cells in non-ionic detergent*

The following method was used in the purification of a major glycoprotein from rat thymocytes (41). In the procedure Nonidet P-40 (NP-40), Lubrol-PX or Triton X-100 could be substituted for the Brij-96. The choice of detergent is based on which gives the best yield of antigenic activity in a high speed supernatant as determined by inhibition of the indirect binding assay (Section 5.1.3).

(i) Tease thymocytes into Tris/140 mM saline buffer pH 8, to give $1.5 \times 10^9$ cells/ml.
(ii) Add 10% (w/v) Brij-96 in Tris/140 mM saline buffer pH 8 plus 10 mM iodoacetamide, 1 mM PMSF (di-isopropyl fluorophosphate used in ref. 41) to give a final Brij-96 concentration of 3%. Homogenize with six strokes of a Potter−Elvehjem homogenizer; stir for 30 min and remove nuclei by centrifugation at 6300 *g* for 20 min.
(iii) Add 3% sodium deoxycholate in Tris buffer pH 8 to the supernatant to give a final concentration of 1% deoxycholate, 2% Brij-96 and centrifuge at 70 000 *g* for 60 min. The supernatant can then be applied to an affinity column or stored at −40°C.

5.2.7 *Solubilization of brain homogenate in deoxycholate*

(i) Take 150 g wet weight of rat brain and homogenize in a Waring blender with

**Table 2.** Purification of MRC OX-45 antigens from spleen and brain endothelium.

| *Fraction* | *Protein (mg)*[a] | *$10^{-4}$ × antigenic activity (units)*[b] | *Yield of antigenic activity (%)* | *Relative specific activity* |
|---|---|---|---|---|
| Spleen | | | | |
| Tween extract from 520 g spleen | 52 000 | 1937 | 100 | 1 |
| Crude membranes | 7700 | 1395 | 72 | 4.9 |
| Deoxycholate extract | 7000 | 620 | 32 | 2.2 |
| Passed through MRC OX-45 Ab column | 7540 | 31 | 1.6 | – |
| Eluted from MRC OX-45 Ab column | – | 434 | 22 | – |
| After Sephacryl S300 chromatography | – | 291 | 15 | – |
| After deoxycholate removal | 1.06 | 437 | 22.6 | 11 000 |
| Brain | | | | |
| Deoxycholate extract from 340 g brain | 17 000 | 41.2 | 100 | 1 |
| Passed through MRC OX-45 column | 16 200 | ~0 | ~0 | – |
| Eluted from MRC OX-45 column | – | 25.6 | 62 | – |
| After Sephacryl S300 chromatography | – | 18.7 | 45 | – |
| After deoxycholate removal | 0.114 | 32.5 | 79 | 110 000 (170 000)[c] |

[a]Protein was assayed by the Lowry method for the extracts and by amino acid analysis for the pure glycoprotein.
[b]One unit of MRC OX-45 antigenic activity is the amount of antigen needed to give 50% inhibition of the assay for antigenic activity.
[c]The overall purification of brain MRC OX-45 antigen was 170 000 with respect to the starting brain homogenate as estimated from the quantitative absorption data.

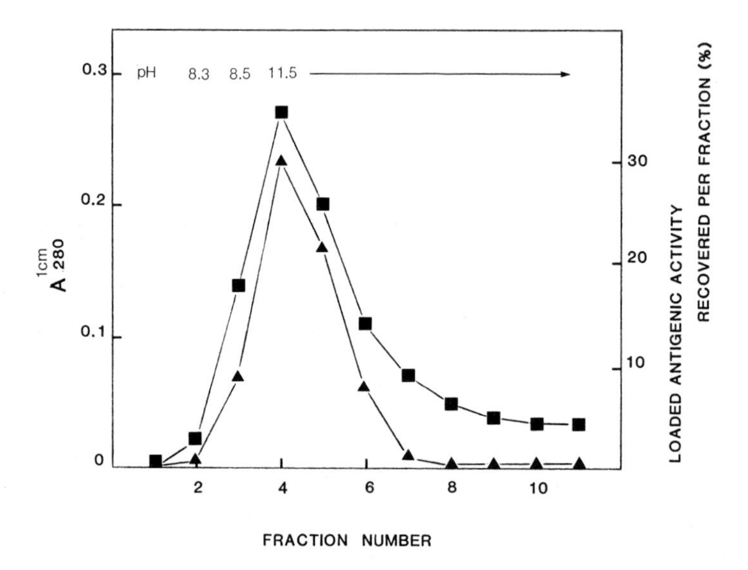


**Figure 7.** Elution of OX-45 spleen antigen from antibody affinity column with 0.05 M diethylamine–HCl, 0.5% sodium deoxycholate, 0.02% $NaN_3$ pH 11.5. 4 ml fractions were collected and assayed for pH, absorbance at 280 nm (■) and antigenic activity (▲).

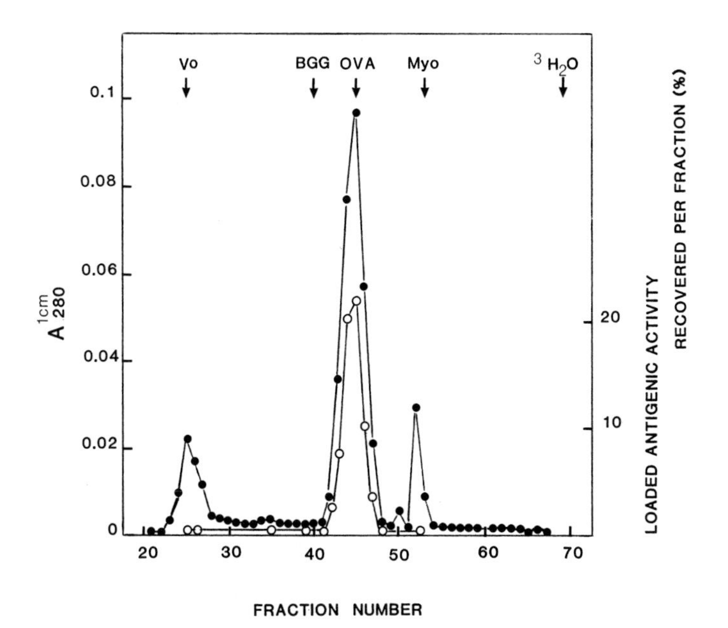


**Figure 8.** Gel filtration of spleen OX-45 antigen. Material eluted from the antibody affinity column was chromatographed on Sephacryl S300 (2.2 × 75 cm) in 0.5% sodium deoxycholate, 0.01 M Tris–HCl, 0.02% $NaN_3$ pH 8. 4.5 ml fractions were collected and assayed for antigenic activity (○) and protein content by absorbance 280 nm reading (●). The elution positions of markers run separately are shown: Vo, void; BGG, bovine γ-globulin; OVA, ovalbumin; Myo, myoglobin.

600 ml of 30 mM Tris–HCl, 0.02% $NaN_3$ pH 8. Centrifuge at 18 000 *g* for 60 min and discard the supernatant.

(ii) Suspend the pellet in 1.25 litres of 3% deoxycholate, 30 mM Tris–HCl, 0.02% $NaN_3$ pH 8, iodoacetamide 1 mM, PMSF 0.1 mM. Briefly homogenize in a Waring blender and then in a Potter–Elvehjem homogenizer as described in Section 5.2.4 above.

(iii) Centrifuge at 18 000 *g* for 60 min and retain the supernatant. Re-homogenize the pellet in 200 ml of 3% deoxycholate buffer and repeat as above.

(iv) The extracts can be immediately used for affinity chromatography or frozen and re-centrifuged before use (this can lead to a less viscous extract).

### 5.2.8 *Purification of the OX-45 antigen from spleen*

(i) Prepare a crude membrane fraction from a Tween-40 homogenate of Sprague–Dawley rat spleens (520 g from about 1000 rats) and then solubilize in sodium deoxycholate as described above (Sections 5.2.4 and 5.2.5).

(ii) Purify antigen from the deoxycholate extract using either the column or batch procedures. Use 10 ml of Sepharose CL-4B with OX-45 IgG coupled at 7 mg IgG/ml beads after pre-elution with a high pH buffer. Both procedures give good results in that antigen from about 1 litre of extract is almost totally depleted and exhibits a similarly high degree of purity (*Table 2*).

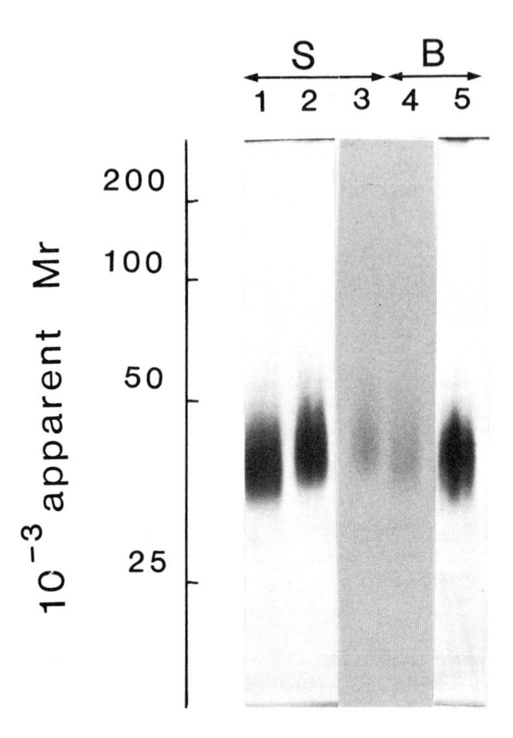


**Figure 9.** Analysis of MRC OX-45 antigens by 10% SDS−PAGE on 10% acrylamide gels. S and B denote whether the antigens were purified from spleen or brain, respectively. The amounts of protein loaded were 1.3 μg for the gels stained with a silver stain **(tracks 1, 2** and **5)** and 5 μg for the gels stained with periodic acid/Schiff **(tracks 3** and **4)**. All samples were reduced with dithiothreitol except **track 1**.

(iii) Wash the columns first with Tris/0.5% deoxycholate buffer, then with 0.15 M NaCl in the above buffer (see Section 4.6) and finally with the Tris/0.5% sodium deoxycholate buffer until the absorbance at 280 nm returns to baseline.

(iv) Elute bound material with 0.05 M diethylamine/HCl, 0.5% deoxycholate, 0.02% $NaN_3$ pH 11.5 and measure the *A*1% 280 nm of eluted fractions prior to neutralization to pH 8−9 with solid glycine. A typical elution profile is shown in *Figure 7* with material beginning to emerge slightly before the pH rises, presumably due to the buffering capacity of the column.

(v) Concentrate the eluted material, dialyse and analyse on SDS−PAGE where small amounts of shed antibody are detected only after the first use of the column.

(vi) The final purification step consists of gel filtration on a Sephacryl S300 column where the antigenic activity migrates as a symmetrical peak coincident with the major peak of protein (*Figure 8*).

(vii) Those fractions containing antigenic activity and of highest purity as assessed by SDS−PAGE are pooled, concentrated and dialysed against 0.1 M $NH_4HCO_3$ to remove deoxycholate prior to biochemical studies. The pure material runs as a rather diffuse band on SDS−PAGE (*Figure 9*) and this is presumably due to the high carbohydrate content.

### 5.2.9 *Purification of OX-45 antigen from brain*

The starting material was a deoxycholate extract from 340 g rat brain that had been prepared as in Section 5.2.7 and used for the purification of OX-2 antigen (42) after which it had been stored at −40°C. On thawing proteolytic inhibitors were added and the extract centrifuged at 70 000 *g* for 60 min.

The yield and purification factor are given in *Table 2* and the pure antigen shown in *Figure 9*. It can be seen that the brain antigen is similar on SDS−PAGE to that from spleen with the small difference in apparent molecular weight due to differences in carbohydrate composition (33). The purification factor from the extract was 110 000-fold and this equals at purification of 170 000-fold from starting brain homogenate. In this case no problems were seen with non-specifically bound material or leakage of antibody from the column even though a very minor component of the brain homogenate was purified in small amounts. The OX-45 from brain is largely derived from the brain capillaries (33).

## 6. REFERENCES


1. Silman,I.H. and Katchalski,E. (1966) *Annu. Rev. Biochem.*, **35**, 873.
2. Porath,J., Axen,R. and Ernback,S. (1967) *Nature*, **215**, 1491.
3. Boegman,R.J. and Crumpton,M.J. (1970) *Biochem. J.*, **120**, 373.
4. Livingston,D.M. (1974) In *Methods in Enzymology*. Jakoby,W.B. and Wilchek,M. (eds), Academic Press, New York, Vol. **34**, p. 723.
5. Anfinsen,C.B., Bose,S., Corley,L. and Gurari-Rotman,D. (1974) *Proc. Natl. Acad. Sci. USA*, **71**, 3139.
6. Letarte-Muirhead,M., Barclay,A.N. and Williams,A.F. (1975) *Biochem. J.*, **151**, 685.
7. Robb,R.J., Strominger,J.L. and Mann,D.L. (1976) *J. Biol. Chem.*, **251**, 5427.
8. Robbins,J.B. and Schneerson,R. (1974) In *Methods in Enzymology*. Jakoby,W.B. and Wilchek,M. (eds), Academic Press, New York, Vol. **34**, p. 703.
9. Kohler,G. and Milstein,C. (1975) *Nature*, **256**, 495.
10. Williams,A.F., Galfre,G. and Milstein,C. (1977) *Cell*, **12**, 663.
11. Sunderland,C.A., McMaster,W.R. and Williams,A.F. (1979) *Eur. J. Immunol.*, **9**, 155.
12. Parham,P. (1979) *J. Biol. Chem.*, **254**, 8709.
13. Secher,D.S. and Burke,D.C. (1980) *Nature*, **285**, 446.
14. Williams,A.F. and Barclay,A.N. (1986) In *Handbook of Experimental Immunology*. Weir,D.M. and Herzenberg,L.A. (eds), Blackwell Scientific Publications Limited, Oxford, 4th edition, Chapter 22.
15. Williams,A.F. (1977) In *Contemporary Topics in Molecular Immunology*. Ada,G.L. and Porter,R.R. (eds), Plenum Press, New York, Vol. **6**, p. 83.
16. Jensenius,J.C. and Williams,A.F. (1974) *Eur. J. Immunol.*, **4**, 91.
17. Mason,D.W. and Williams,A.F. (1980) *Biochem. J.*, **187**, 1.
18. Mason,D.W. and Williams,A.F. (1986) In *Handbook of Experimental Immunology*. Weir,D.M. and Herzenberg,L.A. (eds), Blackwell Scientific Publications Limited, Oxford, 4th edition, Chapter 38.
19. Toda,K., Bozzaro,S., Lottspeich,F., Merkl,R. and Gerisch,G. (1984) *Eur. J. Biochem.*, **140**, 73.
20. Wilchek,M., Mikon,T. and Kohn,J. (1984) In *Methods in Enzymology*. Jakoby,W.B. (ed.), Academic Press, New York, Vol. **104**, p. 3.
21. Porath,J. (1974) In *Methods in Enzymology*. Jakoby,W.B. and Wilchek,M. (eds), Academic Press, New York, Vol. **34**, p. 13.
22. Schneider,C., Newman,R.A., Sutherland,D.R., Asser,U. and Greaves,M.F. (1982) *J. Biol. Chem.*, **257**, 10766.
23. Read,R.S.D., Cox,J.C., Ward,H.A. and Nairn,R.C. (1974) *Immunochemistry*, **11**, 819.
24. Wilchek,M., Bocchini,V., Becker,M. and Givol,D. (1971) *Biochemistry*, **10**, 2828.
25. Edgington,T.S. (1971) *J. Immunol.*, **106**, 673.
26. Gee,N.S. and Kenny,A.J. (1985) *Biochem. J.*, **230**, 753.
27. van Oss,C.J., Absolom,D.R., Grossberg,A.L. and Neumann,A.W. (1979) *Immunol. Commun.*, **8**, 11.
28. Herrmann,S.H., Ming Chow,C. and Mescher,M.F. (1982) *J. Biol. Chem.*, **257**, 14181.

29. Liebman,H.A., Limentani,S.-A., Furie,B.C. and Furie,B. (1985) *Proc. Natl. Acad. Sci. USA*, **82**, 3879.
30. Jefferies,W. (1985) Ph.D. Thesis, University of Oxford.
31. Unkeless,J.C. (1979) *J. Exp. Med.*, **150**, 580.
32. Hsiung,L.-M., Barclay,A.N., Brandon,M.R., Sim,E. and Porter,R.R. (1982) *Biochem. J.*, **203**, 293.
33. Arvieux,J., Willis,A.C. and Williams,A.F. (1986) *Mol. Immunol.*, **23**, 983–990.
34. Towbin,H., Staehelin,T. and Gordon,J. (1979) *Proc. Natl. Acad. Sci. USA*, **76**, 4350.
35. Heusser,C.H., Stocker,J.W. and Gisler,R.H. (1981) In *Methods in Enzymology*. Langone,J.J. and Vunakis,H.V. (eds), Academic Press, New York, Vol. **73**, p. 406.
36. Barclay,A.N. (1977) *Brain Res.*, **133**, 139.
37. Helenius,A. and Simons,K. (1975) *Biochim. Biophys. Acta*, **415**, 29.
38. Johnson,P., Gagnon,J., Barclay,A.N. and Williams,A.F. (1985) *EMBO J.*, **4**, 2539.
39. Fulton,R.J. and Hart,D.A. (1981) *Biochim. Biophys. Acta*, **642**, 345.
40. Standring,R. and Williams,A.F. (1978) *Biochim. Biophys. Acta*, **508**, 85.
41. Brown,W.R.A., Barclay,A.N., Sunderland,C.A. and Williams,A.F. (1981) *Nature*, **289**, 456.
42. Barclay,A.N. and Ward,H.A. (1982) *Eur. J. Biochem.*, **129**, 447.

CHAPTER 6

# Gel immunodiffusion, immunoelectrophoresis and immunostaining methods

DAVID CATTY and CHANDRA RAYKUNDALIA

## 1. INTRODUCTION

This chapter covers methods of immunodiffusion and immunoelectrophoresis with agar and agarose gels and the immunostaining of Western blots of polyacrylamide gels.

Immunoprecipitation in gels is the basis of a number of important techniques widely applied in the study of antibodies and for the detection and measurement of soluble antigens. It depends upon the simple principle that gels support an aqueous phase through which most macromolecules ($>10^6$ molecular weight) will diffuse freely. When multi-determinant complex antigens diffuse into a zone of antibody, conditions are met where the proportions of the intercombining components are optimal for visible precipitation. Reactions are conventionally performed in thin horizontal beds of agar supported on glass plates in which antigens and antibodies are placed in opposing wells to produce single and double diffusion reactions (1,2). Modifications of this simple diffusion principle include an initial electrophoretic separation of antigens by charge, followed by a diffusion step [immunoelectrophoresis, IEP (3)], electrophoresis of antigens into antibody-containing agarose [rocket and two-dimensional immunoelectrophoresis, RIEP and 2D-IEP, respectively (4–6)] which accelerates precipitation and allows quantitation and better differentiation of antigens, and diffusion of antigen into an antibody-agar (or the reverse) which gives a quantitative radial precipitation result (7,8). We give protocols for performing these tests, some examples of their many applications and consider their respective advantages and disadvantages. The last section describes the immunostaining of antigen bands obtained by the Western blot method (9,10) following separation by polyacrylamide gel electrophoresis (PAGE). This is an important technique for defining the molecular size of antigens and the heterogeneity of antigens and antibodies.

## 2. EQUIPMENT AND MATERIALS FOR GEL PRECIPITATION TESTS

### 2.1 **Major items** (see Appendix for suppliers)

Electrophoresis power pack: multi-channel needed for use of several boxes.
Electrophoresis box: large buffer reservoir version is preferable with water-flow cooling plate.
Water bath.
Horizontal table or levelling board (screw adjustable corner supports) and spirit level.
Electric warming plate (to 60°C).
Magnetic stirrer hot plate.

2.2 **Small items** (see Appendix for suppliers)

Glass plates 8 × 8 × 0.15 cm.

Gel punches—useful range is 1.5 mm (2 μl), 2.3 mm (5 μl), 3.0 mm (10 μl), 5 mm (30 μl), 6 mm (40 μl) (volumes relate to 1.5 mm gels). Drawn, square-cut glass pipette tips can be used or a set of cork borers. Manufactured ranges are also available.

Tap Venturi suction pump with tip (Pasteur pipette) for removing gel plugs.

Pointed scalpel blade with handle.

Secotome blade (>8 cm).

Metal ruler and two 10 × 1 × 1 cm support bars for cutting troughs.

Various template designs of wells and troughs on cards (*Figures 2, 7, 10, 13, 15, 16* are examples).

Micropipette with fine bore tips. Disposable tip models preferred to avoid sample contamination.

Variable micropipette, 10–250 μl range, with tips.

Micrometer eye piece for measuring ring precipitations.

Whatman No. 1 filter papers.

Paper towels.

Gelbond film squares 8 × 8 cm as gel supports.

Metal container (small saucepan) for boiling agar gel.

Clear plastic storage and staining boxes (10 × 10 × 2 cm) with lids.

Pyrex bottles (50 ml) with caps (or 20 ml glass Universals) for prepared agar gel.

Pyrex conical flask (2 litre) for preparing bulk agar solution.

Graduated pipettes (10 ml) with bulb or pipette aid (e.g. Pi-pump) (preferred).

Glass fine-tip Pasteur pipettes and teats.

Glass test tubes (15 ml) and rack (for water bath).

Test tubes (5 ml) for sample dilution.

Surgical lint for electrophoresis wicks.

Parafilm.

*Figure 1* shows a bench layout of equipment and items used for gel tests.

2.3 **Chemicals, buffers and reagents**

Agarose—low electroendosmosis grade is required for most electrophoresis applications and can be used for diffusion reactions.

Agar—for gel double diffusion (GDD) and radial immunodiffusion tests. Some brands of agar can be substituted for more expensive agarose (see Appendix).

Barbitone buffer: 0.05 M, pH 8.6; 9.21 g of barbitol (5,5-diethyl barbituric acid) in 3 litres of distilled water. Add 51.54 g of sodium barbitone. Mix and when dissolved add 5 g of sodium azide and make up to 5 litres.

Phosphate-buffered saline (PBS), pH 7.2, 0.1 M $PO_4$.

$Na_2HPO_4.12H_2O$ – 2.7 g
$NaH_2PO_4.2H_2O$ – 0.39 g
NaCl – 9.0 g
} dissolve in 1 litre of distilled water.

Saline—0.85% (w/v) NaCl in distilled water.

Protein staining solution and destainer:

(i) Mix distilled water, glacial acetic acid and methanol in proportions 5:1:5 to make 2.5 litres.

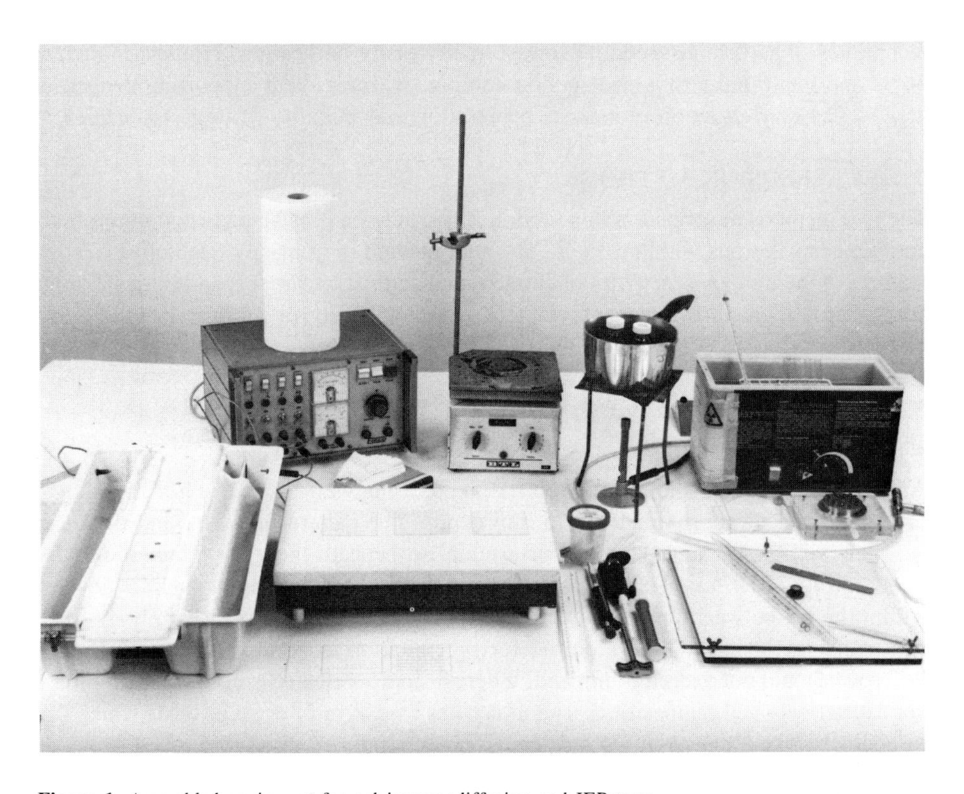

**Figure 1.** Assembled equipment for gel immunodiffusion and IEP tests.

(ii) To 1 litre add 5 g of Coomassie Brilliant Blue R (stain).
(iii) Remaining 1.5 litres is the destainer.

Polyethylene glycol (PEG) (6000 mol. wt).
Bromophenol Blue crystals.
Standard antigens at 1 mg/ml (individual antigens). In some tests for serum proteins a pool of normal serum can be used. For human serum proteins this can be calibrated against a WHO serum standard with known component concentrations (see Appendix).
Reference antisera—unispecific or polyspecific according to test (see Appendix for suppliers).

## 2.4 Preparation of 1% (w/v) agarose–barbitone gel

(i) Add 5 g of agarose powder to 500 ml of hot barbitone buffer in a conical flask (2 litre) and heat/stir on a hot plate magnetic stirrer.
(ii) When dissolved completely and clear, add 15 g of PEG and continue to stir until dissolved.
(iii) Dispense into 20–40 ml volumes in bottles, cap and store at 4°C.

Agar, as opposed to agarose, can also be prepared at 1.0–1.5% (w/v) in PBS or saline with 3% PEG for immunodiffusion and some IEP tests [see Section 4.1.4 (ii)]. It is convenient, however, to prepare a standard agarose gel which can be used for all techniques. Agarose has good solubility, melts easily on heating, remains molten at 56°C and sets rapidly to a clear gel on cooling, with excellent supporting properties at 1%. Gel concentration influences penetration and mobility of large molecules.

## 3. GEL DIFFUSION METHODS

The two methods described in this section rely solely on passive diffusion of antibody and antigen solutions within the gel. The first method is primarily qualitative, testing features of the immunospecificity of antigens and antibodies; the second is a quantitative adaptation which allows measurement of either antibody or antigen in unispecific systems.

### 3.1 **Gel double diffusion (GDD)**

#### 3.1.1 *Principle*

Antigen and antibody solutions are placed in opposing wells cut into a horizontal agar or agarose gel of approximately 1.5 mm depth. The pattern of wells, and distances between wells, is arranged by use of a template beneath the glass. This allows for standardization. Diffusion occurs radially from the wells and precipitation lines develop within the gel between opposing wells. The gap between antigen and antibody wells is a major factor in the time required for precipitation to occur. As a general guide the space should not exceed 6 mm and, where a single well size is used, the gap should not exceed two well diameters. Large molecules ($>10^6$ mol. wt, e.g. IgM) which diffuse slowly take longer than 24 h to develop. Generally reactions are well developed after overnight storage at 4°C. Increasing temperature accelerates reactions but sometimes with a loss of definition. Well sizes can be adjusted to give optimal proportions of reagents and sharply defined lines in the intervening gel. Poorly adjusted relative concentrations may give no precipitation, as complexes remain soluble, or fuzzy and blurred lines. Ideal reactions may be obtained by using several concentrations (dilutions) of antigen and antibody. As a general guide for undiluted antisera, antigens should be approximately 1 mg/ml. Antisera with good precipitating properties used at appropriate dilutions may detect single antigens at 5 $\mu$g/ml. *Figure 2* illustrates some useful well patterns for GDD analysis. These are actual size and can be used as templates. Gel punches with fixed ring patterns can be purchased from some suppliers (see Appendix).

#### 3.1.2 *Applications*

(i) An essential initial test for precipitating properties, approximate titre and specificity of antisera raised to antigens in solution.
(ii) A standard method for determining specificity of antisera.
(iii) A test for antigenic identity, determining the purity of antigen, and antigenic relationships between molecules.
(iv) An approximate test of antigen concentration.

#### 3.1.3 *Preparing gel plates*

(i) Loosen the cap on a bottle of stored agarose and stand the bottle in boiling water until the agarose is completely molten.

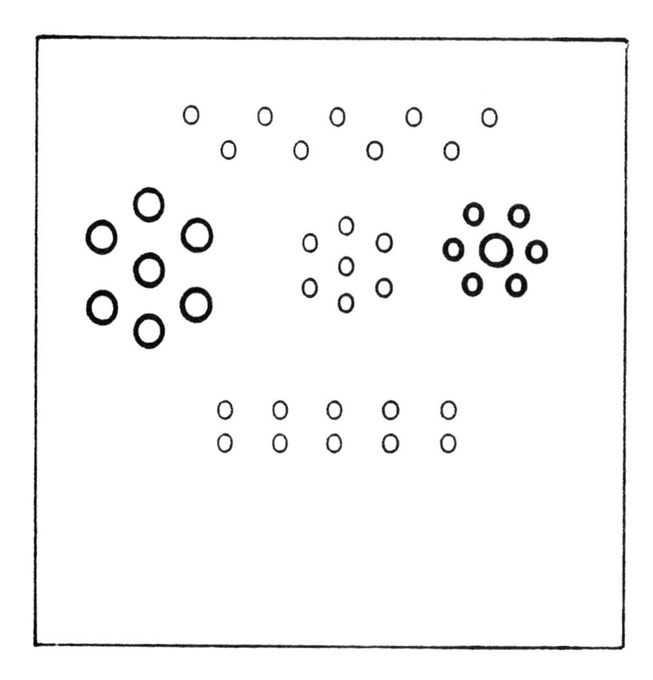

**Figure 2.** Template for GDD reactions. Patterns can be altered to suit the particular needs of the test. The patterns are actual size. A photocopy made of this figure stuck to a card forms a working template. This applies also to *Figures 7, 10, 13, 15* and *16*.

(ii) If using a Gelbond support, wet the square and stick the hydrophobic side to a glass plate. Put the glass plate onto a level board or table top and check that the plate is level.

(iii) Pipette 9.6 ml of molten agarose onto the plate allowing this to run from the centre to the edges. Use the tip of a pipette to assist this if necessary, ensuring no bubbles remain in the gel. Lumpy and cloudy gels indicate that the agarose was insufficiently heated.

(iv) Allow the gel to set and then place it on a template and cut out the well patterns. The plugs can be removed by suction or gently lifted out with a pointed scalpel blade or hypodermic needle (a hooked tip is useful).

### 3.1.4 *Performing and preserving gel tests*

(i) Fill the wells with antigens and antibodies using micropipettes with tips suitable for the well size. Be careful not to dislodge the gel or overflow the wells. Wells should be approximately filled and appropriate well sizes selected for this.

(ii) Place the gel in a storage box on a dampened filter paper and apply the cover. Never leave gels exposed on the open bench as they will dry rapidly. Gels can be left to develop at room temperature, but if an overnight period is required they are best stored at 4°C. Gels should be examined repeatedly. Strong reactions may begin to develop within a few hours. When optimal (up to 24 h), gel reactions can be stopped by washing and the results preserved by drying and staining.

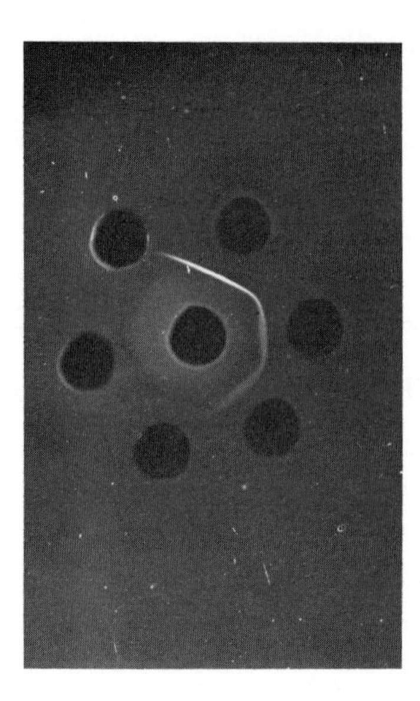

**Figure 3.** Gel immunodiffusion reaction in a ring pattern to show how ratios of antibody:antigen critically influence the position and sharpness of the precipitation line. The antigen (human IgG at 1 mg/ml) is in the centre well; anti-IgG is diluted from 1:20 (outer top right well) in doubling dilutions clockwise around the antigen well.

(iii) Immerse the gel in saline in a storage box and soak for several hours or preferably overnight. Use of Gelbond prevents the agar from coming free from the support. Change to distilled water and soak for several hours and then remove the plate to the bench.

(iv) Dry the gel by applying water-dampened filter paper onto the gel surface and add several layers of paper towel. Place a book or some article of equivalent weight ($\sim 1$ kg) on the top and allow the gel to press-dry. After 2 h remove weight and wet papers, leaving the filter paper to dry out. A hair drier can be used to assist drying. When the paper dries it will lift off to leave a clear dry film of agar on the Gelbond or glass. The Gelbond can be lifted off the glass at this stage.

(v) Stain the gel by placing the glass/agar or Gelbond/agar into Coomassie Blue R stain for about 1 min. It is helpful to use a storge box and hold this above a lamp to observe the stain development. Return the stain to the bottle and wash the plate under the tap. Remove the background staining in destainer, wash the plate again and allow it to dry. Gelbond plates can be conveniently mounted in albums. Glass plates can be card-mounted and filed or covered with broad Sellotape and stored upright in boxes.

### 3.1.5 *Interpretation of gel reactions*

(i) *The importance of balanced concentrations—use of antibody:antigen ratios for titration. Figure 3* illustrates how differences in concentration of antigen (outer wells, human IgG) affect the position of the precipitation line with antibody (centre well, anti-

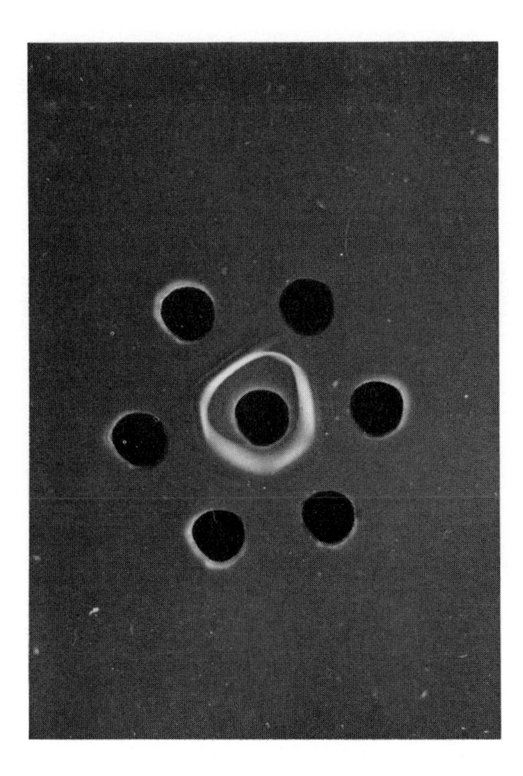

**Figure 4.** GDD test to show the presence of a pure antigen in one well (**top right**) and contamination with other antigen preparations tested in other wells prior to satisfactory purification. Reactions of identity are seen with the same antigen in all wells, although the proportions of this antigen to antibody in the centre well does not give a sharp precipitation line in some cases. The antiserum in the centre well is an anti-whole human serum and the purified antigen is albumin.

human IgG), and change the definition of the reaction. This emphasizes the point that in testing both antisera and antigens, a range of dilutions of reagents should be applied. In multiple precipitation systems not all lines may be seen at one set of dilutions. This principle can be applied, with standard antigen, to initial tests on the precipitating properties and titre of sera. A good antiserum will give a sharply defined precipitation line mid-distance between the antigen and antibody wells at dilutions greater than 1:50.

(ii) *Multiple antigen–antibody systems and tests for purity of antigen. Figure 4* illustrates a multispecific antiserum (centre well) reacting with a mixed antigen solution (well 1) and an antigen at increasing degrees of purity. Continuous lines of precipitation between the antibody well and outer wells are reactions of complete identity, indicating the presence of the same antigens in the different preparations (i.e. molecules with exact antigenic identity).

(iii) *Demonstration of antigens with complete, partial- and non-identity. Figure 5* illustrates how, by using a polyspecific antiserum in the centre well, the antigenic relationship of molecules placed in the outer wells can be determined. Two non-cross-reacting antigen systems are present; IgG in wells 1, 2 and 3, and serum albumin in wells 3, 4 and 5. Each antigen precipitates with its respective antibody diffusing from

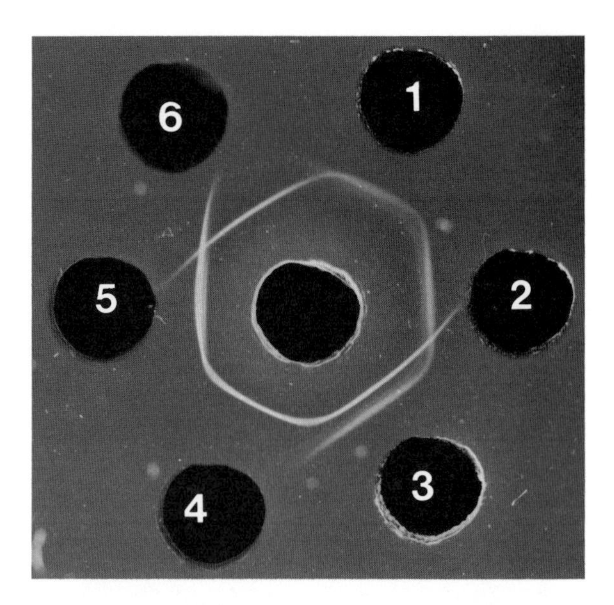


**Figure 5.** GDD test to show a reaction of partial identity between IgG whole molecule (**well 1**) and F(ab')$_2$ (**well 6**) with antibodies (**centre well**) to intact IgG. The position of the spur indicates that intact IgG has epitopes not shared with F(ab')$_2$ to which the antibodies can bind to extend the precipitation line. Also shown are reactions of non identity between human serum albumin (**well 5**) and F(ab')$_2$ (**well 6**), and between albumin (**well 3**, mixed with IgG) and IgG (**well 2**). Reactions of complete identity are seen between the IgG in **wells 1, 2** and **3**, and between albumin in **wells 3, 4** and **5**. The centre well contains antibodies to both IgG and albumin and the two precipitation lines against well 3 identify the presence of both antigens.

the centre well, and reactions of complete identity have developed where the same antigen exists in adjacent wells. The two antigen precipitation systems cross over between wells 2 and 3 in a reaction of non-identity as no epitopes are shared between these two antigens. In well 6 is the F(ab')$_2$ fragment of IgG which lacks epitopes of the Fc region of the intact molecule. This produces a reaction of partial identity with the IgG of well 1, evident from the precipitation 'spur' that extends outwards from the main reaction line; the spur is in the direction of the incomplete antigen F(ab')$_2$ (in relation to available antibody specificities). Antibodies diffusing towards well 6 and meeting F(ab')$_2$ have some (anti-Fc) specificities not satisfied by the restricted range of F(ab')$_2$ epitopes. These will continue to diffuse until they meet intact IgG molecules with the extra Fc epitopes to which they bind and thereby extend the IgG precipitation outwards as a spur. There is a reaction of non-identity between F(ab')$_2$ of IgG (well 6) and albumin (well 5).

(iv) *The specificity of antibodies. Figure 6* shows a simple test in which a reference antiserum (anti-Y) to one antigen (well 5) is compared with two test antisera (wells 3 and 4). The antigen system in the centre well is a mixture of molecules containing (at least) two components, one of which is the target antigen (Y) for the reference antiserum. Standard purified antigen Y is in well 6 and there is a reaction of identity between reference anti-Y, standard antigen Y and antigen Y in the centre well. Test antibody 1 (well 4) is also specific to antigen Y and extends the complete identity reaction with no further lines. However, the test antibody 2 (well 3) contains antibodies to a second antigen (antigen X) in the centre well. The identity of the second antigen is

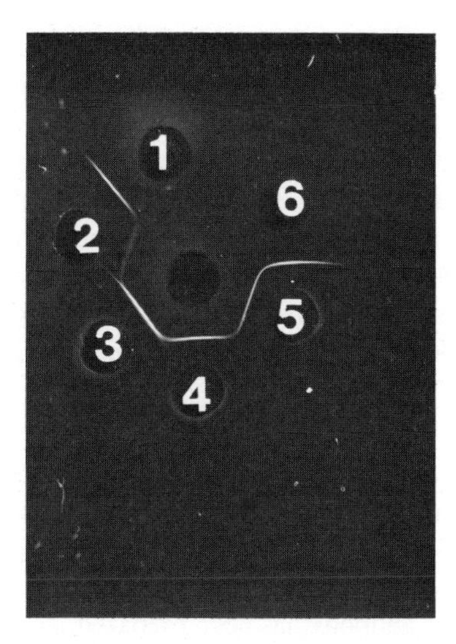


**Figure 6.** GDD reaction to test the specificity of antisera and the importance of using reference antibodies and standard antigens to ensure a correct interpretation. See text for details.

confirmed by the reference anti-X serum in well 2 and this reaction gives complete identity with purified standard antigen X in well 1. The two antigen reactions coincide against well 3 but are independent systems that emerge between wells 2 and 3, and wells 3 and 4. They are not reactions of partial identity with spurs, as might be deduced without the use of standard antigens and reference antibodies.

#### 3.1.6 *Advantages and disadvantages*

Gel double diffusion stands alone as a unique and simple method for studying molecular antigenicity and the specificity of antibodies. It has two major disadvantages, (i) it has relatively low sensitivity when compared with indirect methods such as agglutination and enzyme assays and (ii) examination of multiple antigen system by GDD is often complicated by overlapping and superimposed precipitation lines. In this respect it is inferior to IEP (Section 4.1) which benefits from an initial electrophoresis step to separate the antigens in gel.

### 3.2 **Radial immunodiffusion (RID)**

#### 3.2.1 *Principle*

Molecules in solution placed in agar wells diffuse radially and combine with complementary antibody or antigen equilibrated in the agar fluid phase. In standard RID the agar contains unispecific antibody and the radially-diffusing antigen collects bound antibody until a stable single precipitation ring is formed, which will not increase in size once the reaction is completed by consumption of all antigen determinants. This takes several days, the time dictated by the antigen molecular size. IgM (mol. wt 900 000) takes at least twice as long as IgG (mol. wt 150 000). In practice, for molecules of IgG size or less, 24 h at 4°C or 16 h at room temperature is adequate to obtain a quantitative curve of antigen standard dilutions. Quantitation is based on the fact that the area of the precipitation rings when fully developed over several days has a linear relationship with the quantity of antigen applied to the wells in equal volume. In practice the square of the diameter of perfect radial rings ($D^2$) gives a close approximation to the area concentration relationship over shorter reaction times. Use of standard antigen dilutions over a 10- to 20-fold concentration range, extending down to about 5 $\mu$g/ml, will provide a standard curve from which the test samples can be measured directly from wet gels.

A similar approach can be adopted for comparing the titre of new antisera against a reference reagent. This is known as reverse radial immunodiffusion (RRID) in which the wells of antiserum dilutions diffuse into antigen–agar.

### 3.2.2 *Applications*

(i) *Measurement of single antigens in solution (RID).* The method is routinely used to measure serum proteins, in particular immunoglobulins in suspected multiple myeloma and immunoglobulin deficiency cases. Complement components can also be measured. More generally any antigenic molecule at a concentration greater than or equal to 5 $\mu$g/ml, to which a specific antibody has been prepared, and which is not much larger than $10^6$ molecular weight, can be measured against a known standard. The standard (and test sample) need not be pure and the concentration can be expressed as units relative to standard.

(ii) *Titration of antisera against a reference reagent (RRID).* This is best applied to measurement of IgG antibodies raised by chronic immunization of animals, when single rings develop against the antigen. However, the test has value in routine screening of patients' sera against some antigens for infections although the sensitivity is not sufficient for many systems.

### 3.2.3 *Method—direct (antigen) and reverse (antibody) tests*

(i) Prepare molten agarose (or agar) as for GDD and place the bottle in a 56°C water bath to cool.

(ii) Place 15 ml glass test tubes in a rack in a water bath to warm.

(iii) Place the glass plates on a warming plate (56°C) with or without Gelbond as required.

(iv) Pipette 9.6 ml of agarose into test tubes (one per plate required).

(v) Add the required volume of antibody or antigen, cover with Parafilm and invert gently several times to mix thoroughly, being careful to avoid causing bubbles.

(vi) Transfer a warm plate to the level board or table and rapidly pour the agarose–reagent mixture onto the plate, if necessary using the lip to spread the fluid to the edges on all sides. Bubbles can be removed by touching them with the bottom of the tube or drawing them to the edge.

(vii) Allow the gel to set, cover in a damp storage box and place at 4°C. Plates can be stored for several days as the agar contains sodium azide. Clean filter paper and distilled water should be used, however, in the storage box.

(viii) Place the gel plate on the template (*Figure 7*) and use the 1.5 mm (2 $\mu$l) punch to cut out the wells. Remove the gel plugs.

(ix) Prepare the range of standard antigen (RID) or reference antiserum (RRID) dilutions and the test sample dilutions in small tubes and place 2 $\mu$l of each in

**Figure 7.** Actual size template for RID tests and RRID tests in gel, using 8 cm × 8 cm glass plates and a 5 × 5 well pattern.

the wells by micropipette, using a separate tip for each sample. A reliable micropipette is essential—pipetting error is the largest contributor to poor precision.

(x) Return the loaded gel plate to the storage box and leave it for 24 h (or overnight).

3.2.4 *Technical notes*

(i) Amounts of antibody or antigen incorporated in the gel must be obtained by trial, balancing against the quantity of standard antigen (or reference antibody) in the wells. As a guide, a good antiserum can be used at 0.5% (v/v) in the agarose, and antigens (for RRID) can be incorporated at about 20 μg/ml agarose (200 μg/plate). Antigen standards for RID should be in the range of 1 mg/ml to 5 μg/ml (2 μg to 10 ng/well). Reference antiserum for RRID should be used in wells from undiluted to 1:10. Test sera can be used undiluted and at 1:3.

(ii) The amount of antibody or antigen mixed with the agar is generally set to produce visible ring precipitations that can be measured on wet plates and with the highest standard (or reference) giving about 1 cm diameter on completion. The sensitivity of the assays can be increased by reducing the agar-incorporated reagent concentration and moving the well-applied reagents to a proportionately lower concentration scale. The reduced amounts of combining reagents give poorly visible rings in wet agar and contrast must be recovered by staining the washed and dried plate. Sensitivity can thus be brought to about 1 μg/ml antigen by this tactic.

(iii) It is essential to include an internal standard antigen or antibody dilution range in each plate which can be used as the comparison points for evaluating test samples. Accuracy is lost in cross-reference between plates due to plate variability.

(iv) At the least, duplicate wells of each standard (reference) dilution must be included to reduce the effects of random errors in pipetting and thus to increase precision. Test samples should also be applied in duplicate and it is preferable to use two dilutions.

(v) The agar gel must be of equal depth across the plate to reduce zonal variation. Minor differences can be compensated for by randomizing the position of the standard (reference) wells.

(vi) Plates should be prepared at least 1 day before use and stored at 4°C to allow full equilibration of the incorporated reagent. They should be brought to room temperature before sample application if bench development is intended.

(vii) Ensure that reactions proceed at an even temperature. Significant fluctuations (i.e. from 4°C to the bench) may lead to multiple rings in single antigen systems as the solubility of complexes varies with temperature.

(viii) Antibodies included in agar in RID should be in the form of an IgG fraction to reduce the amount of incorporated protein that gives background staining. A high titre antibody that can be used at concentrations less than 1% (v/v) in agar is desirable for the same reason.

(ix) With a unispecific antibody, RID tests for antigen do not require the antigen source for standards or test samples to be pure. Measurement of many serum proteins is possible using whole serum as the standard for single antigens, and patients' sera as test samples. Likewise the titration of antisera in RRID does not require purified antigen in the gel providing the test sera are known to be unispecific.

(x) Standard microscope slides can be used for RID for small numbers of samples. Agarose and the mixed reagent are used at one-third of the volumes for larger, 8 × 8 cm plates.

### 3.2.5 *Quantitation*

(i) *Reading wet and stained plates.* In many cases wet plates can be read directly. This is done using a micrometer eye piece placed on the reverse side of the glass and holding the plate obliquely above an angle-poise lamp, looking through the plate at a dark background. Plate readers with a diffused light box and graticule lens are also available.

Plates can also be dried and stained as described for GDD (Section 3.1.4) and the reactions then measured. Careful handling should not distort the rings in the drying stage. Stained reactions can be measured with the eye piece or by use of graduated intercepting rulers supplied by some agencies marketing prepared RID plates.

(ii) *Standard curves.* It is only when reactions are allowed to proceed to completion over several days that the ring area assumes a linear relationship to antigen concentration. A compromise that gives good linearity after 24 h is the plot of log antigen concentration against ring diameter (*Figure 8a*). However, popular alternatives are to plot the square of the ring diameter on an arithmetic scale against doubling dilutions of standard (*Figure 8b*) or reducing increments of standard (*Figure 8c*). All three methods of standard curve preparation give results of equivalent accuracy for test samples.

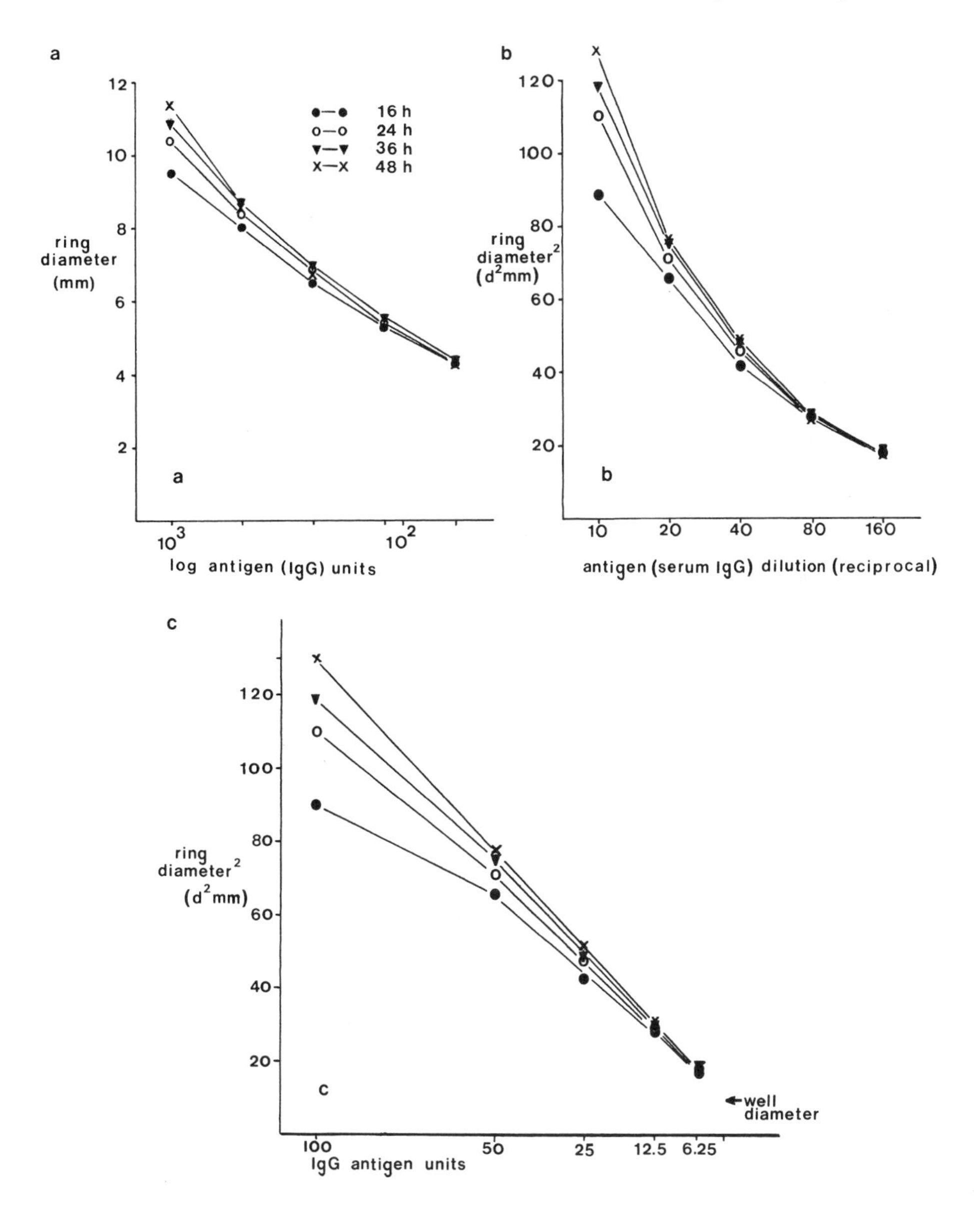


**Figure 8.** Three methods of preparing standard curves of the same data from an RID assay for antigen. The curves in each figure relate to actual ring diameter reading from an assay for human IgG taken at 16, 24, 36 and 48 h. Note that the best linearity is achieved with method **c** but this requires 48 h development. 16 h readings of test samples can be taken from standard curves prepared as in methods **a** and **b** with satisfactory accuracy. The RID plate from which this data is derived is shown as Figure 9, the legend to this giving the conditions of the test.

The data in *Figure 8* derive from a set of standard human IgG concentrations measured at 16, 24, 36 and 48 h on a plate prepared to measure IgG in test human sera. The developed plate is shown as *Figure 9*. Conditions of the assay are given in the footnote. An example of an RRID test of sheep antisera to human IgG is shown in Chapter 2, Figure 8.

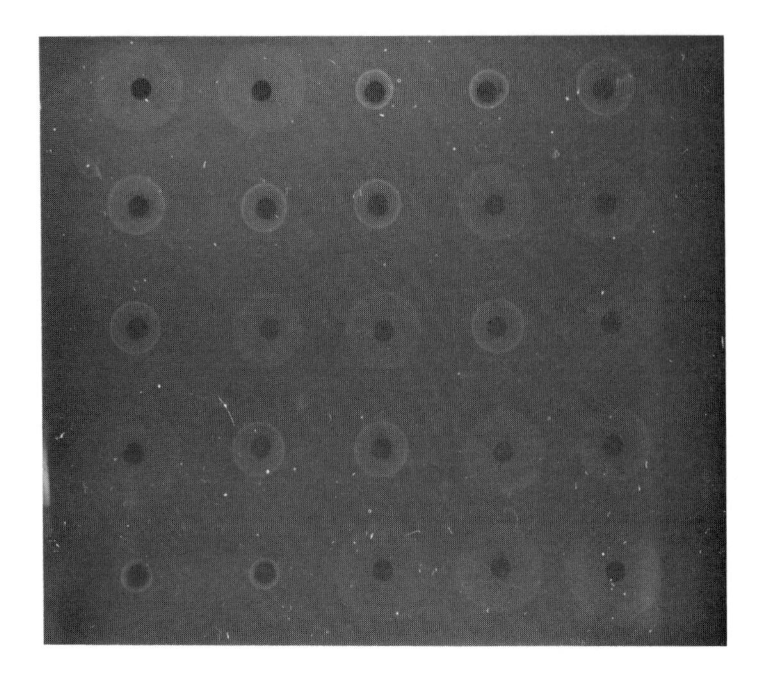

**Figure 9.** RID test for the determination of IgG concentrations in human serum. The agarose contains 100 μl of the IgG fraction of sheep anti-human IgG (Fc-specific) in 9.6 ml. For the standard curve normal human serum pool (of known IgG concentration) has been diluted 1:10; 1:20; 1:40; 1:80 and 1:160, these dilutions being applied to duplicate wells randomly distributed in the plate. The unknowns, which include IgG myeloma sera, are diluted at 1:50 and 1:100 and are tested in duplicate wells. The plate has been developed for 48 h at room temperature.

(iii) *Precision, accuracy and sensitivity.* The precision, or reproducibility of data, in radial diffusion tests is greatly enhanced by using an internal standard curve. Random errors are then confined to intrinsic plate factors of which sample dilution (pipetting) and dispensing inaccuracy are the main contributors. With practice a precision of 5% can be achieved. This can be tested by determining ring diameter variance of duplicate wells and is revealed by irregularities in the derived standard curve. For greatest accuracy, that is closeness of estimate to real values, test sample readings should be contrived, by appropriate dilution, to fall around the centre of the standard curve. With two test dilutions the most central result should be used for calculation and is the rationale behind using two dilutions. Accuracy of readings varies at different sample dilutions because of bias in the curve. In samples lying outside the normal range where neither dilution falls centrally, the test should be repeated with more appropriate dilutions. Accuracy is also increased with the number of replicates of each test sample dilution (by reducing effects of random errors), and by increasing the closeness and range of standard concentrations used to construct the curve.

Sensitivity has two meanings—one relates to the capacity of the standard curve to distinguish differences in test concentration; this is affected by the degree of error in points on the curve and the linearity and slope of the curve. The second relates to the minimum detectable amount of antigen or antibody that can be determined. In RID

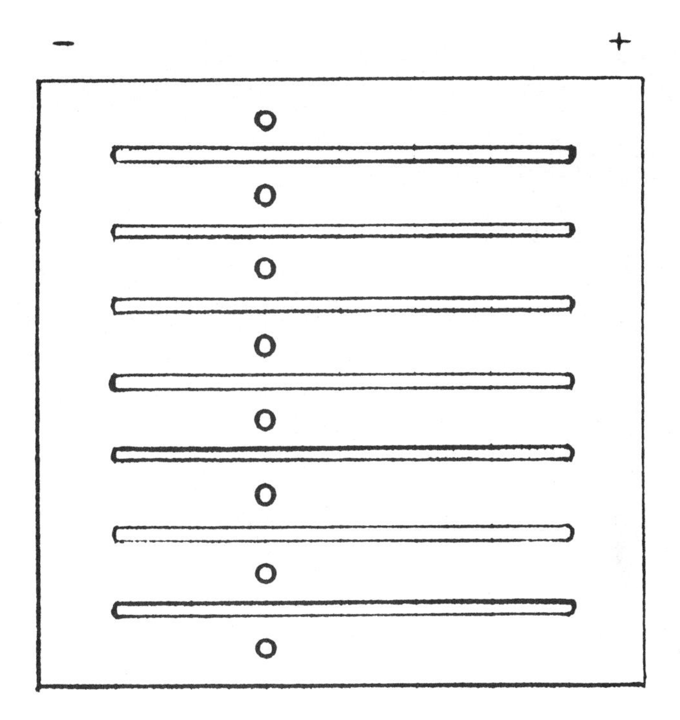


**Figure 10.** Actual size template of wells and troughs for performing immunoelectrophoresis on 8 cm × 8 cm glass plates. The position of the wells shown here is suitable for use of low electroendosmosis agarose and when most antigens migrate anodally (e.g. serum); the position of the wells can be moved to the centre when using other gels of high electroendosmosis character for study of both cathodic and anodic antigens.

this is dictated by well size, random error and visibility and sharpness of precipitation rings at low dilutions.

### 3.2.6 *Advantages and disadvantages of radial immunodiffusion assays*

RID assays are easy to perform, require a minimum of equipment and provide an easily standardizable quantitative system for both antigen and antibodies. The sensitivity is not high ($\sim 5$ $\mu$g/ml) compared with haemagglutination and enzyme assays (Chapter 7 and Volume II) but RID is an appropriate and conventional system for many antigens that come within this concentration range. It demands unispecific precipitating antisera for both RID and RRID, but antigen does not need to be pure for standard or test samples. The assay does not detect abnormalities in antigen (i.e. in structure, restricted antigenicity, homogeneity) but antigen tests are not restricted by pI (cf. RIEP and 2D-IEP, see Section 4). The main disadvantage—other than sensitivity—is the time required for a quantitative result. Data (i.e. reaction plates) can be conveniently stored as permanent records.

## 4. IMMUNOELECTROPHORETIC METHODS

These methods combine the electrophoretic migration of antigens in agarose with immunoprecipitation in gel. The electrophoretic component offers two advantages: (i)

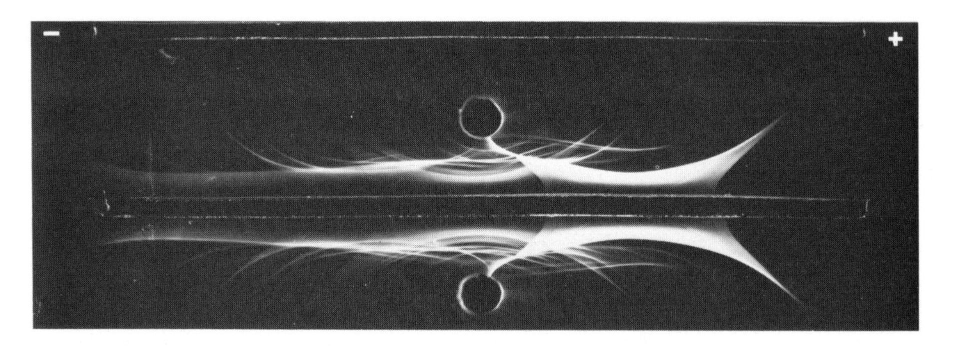


**Figure 11.** Immunoelectrophoresis of guinea pig serum against a polyspecific rabbit antiserum, performed using agar with high electroendosmosis properties.

it achieves a partial separation of mixed antigens, as the rate of their migration in gel is controlled by their different overall charge at the ambient buffer pH, and (ii) it allows negatively-charged molecules to migrate rapidly into an antibody gel to accelerate the precipitation step with minimum lateral diffusion.

There are four popular and widely applicable versions of the immunoelectrophoretic principle. They offer unrivalled facilities for the quantitative and/or qualitative assessment of antigens as single and mixed molecules and are an essential feature of protocols for testing the specificity of antisera.

## 4.1 **Immunoelectrophoresis (IEP)**

### 4.1.1 *Principle*

A pattern of wells and intervening troughs is cut in an agarose plate (*Figure 10*). Antigen solutions are placed in the wells and the agarose is connected by wicks to two buffer reservoirs with electrodes on opposite sides of the plate. Current is applied to move and separate the constituent antigens horizontally along the gel. On completion antisera are placed in troughs parallel to the separated antigen tracks and diffusion of reagents occurs. Where antigen and antibody interact in the agarose, precipitation arcs are formed which define the boundary of discrete antigens (*Figure 11*). Their identity can be determined using known antigens and antibodies (*Figure 12A*), the interpretation of intercepting precipitation arcs follows the principle of gel diffusion methods.

### 4.1.2 *Applications*

IEP is primarily a qualitative method for the study of antigens and of antibody specificity.

(i) *Antigens.* Tests for (a) complexity of antigen solutions—this is of particular value in antigen purification as it identifies the nature of contaminants and (b) analysis of heterogeneity and homogeneity of individual antigens. This method is the 'cornerstone of clinical paraprotein analysis (*Figure 12B*) and a standard method for immunochemical analysis of a wide variety of proteins' (11). It has broad applications in the investigation of antigens of infective agents, plasma proteins and other body fluids of man and animals. It is a useful first screening test for investigating complement and immunoglobulin deficiencies.

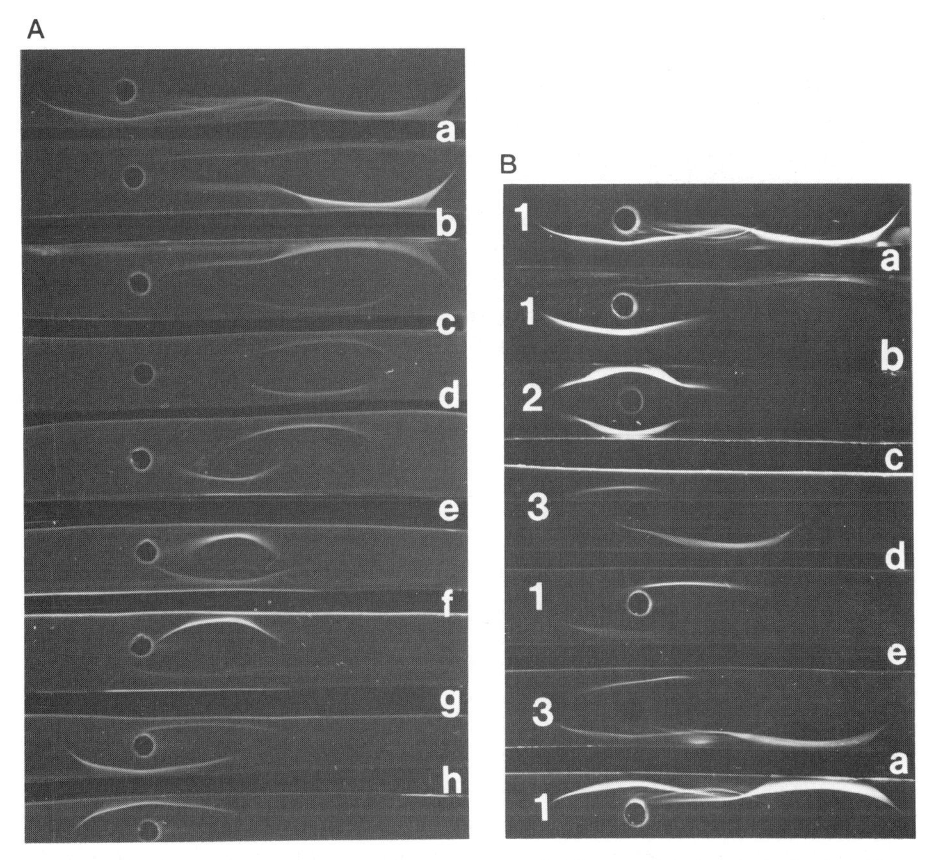


**Figure 12.** Examples of immunoelectrophoresis reactions. (**A**) The use of a selected set of unispecific antisera to demonstrate the electrophoretic position of a number of human serum proteins. All the wells contain normal human serum. The specificity of antisera in the troughs is against: (**a**) whole serum proteins; (**b**) serum albumin; (**c**) $\alpha$-1 anti-trypsin; (**d**) orosomucoid; (**e**) complement C3 component; (**f**) transferrin; (**g**) IgA; (**h**) IgG. (**B**) Demonstration of an IgG$\kappa$ and and IgA$\lambda$ myeloma protein in the serum of two patients, using appropriate antisera. **Wells**: (**1**) normal human serum; (**2**) IgG$\kappa$ myeloma serum; (**3**) IgA$\lambda$ myeloma serum. **Troughs:** (**a**) anti-whole human serum; (**b**) anti-IgG (Fc-specific); (**c**) anti-$\kappa$ light chain; (**d**) anti-IgA (Fc-specific); (**e**) anti-$\lambda$ light chain. Myeloma proteins in serum can be detected by the 'bowing' effect of the high concentration of homogeneous protein on the precipitation arc of the affected immunoglobulin class, and by the fact that the homogeneity of the light chain concentrated in the 'bow' region of the arc results in a restricted light chain isotype staining pattern in the region of the myeloma protein.

(ii) *Antibodies.* A common test for specificity of antisera with unispecific and polyspecific properties (*Figure 12A*). Anti-whole serum proteins are particularly useful reference reagents.

4.1.3 *Method*

(i) Prepare agarose plates as for gel diffusion (Section 3.1.3).

(ii) Apply to a template (*Figure 10*), place supports on each side and, with a metal ruler as a guide, use the pointed scalpel to cut the trough edges. Cut the wells with the 1.5 mm or 2.3 mm punch and cut the ends of the troughs with the 2.3 mm punch. Various designs of IEP gel cutters can be purchased (see Appendix). Remove the well plugs. Invert the plate and at eye level use the scalpel blade to lift one end of each trough gel away from the plate until the gel strip peels away.

(iii) Fill the electrophoresis box reservoirs with buffer and turn on the water supply to the cooling plate.

(iv) Place antigens in the wells of the plate and a small crystal of Bromophenol Blue on the gel surface above one well. Cut off the top right hand corner triangle of gel to denote top and anode side [the glass (or Gelbond) can be etched with a code number]. Place the gel on the box cooling plate (with the direction of the troughs from buffer to buffer). Prepare two clean lint wicks of 25 × 8 cm, soak in buffer, squeeze out the excess and apply the shorter sides to the two edges of the gel to form buffer wicks connecting the buffer reservoirs across the gel. Press gently along the gel edges with fingers dipped in buffer to achieve a continuous connection.

(v) Place the lid on the box (modern box lids connect the circuitry for safety). Connect the terminals to the box ensuring correct polarity [red to red = anode (+) to right side of plate]. Check the terminal polarity on the power pack and switch on the power supply. Adjust to approximately 80 V across the plate (10 V/cm) or 12 mA per plate (1.5 mA/cm). Large electrophoresis boxes will run up to five gels concurrently at a setting of about 150 V and 60 mA. Running gels at too high a voltage can lead to overheating, distortion, drying of the gel and loss of current across the plate.

(vi) Check the migration (anodal) of the Bromophenol Blue. At pH 8.6 this should run at about 4 cm/h. Always turn off the power before lifting the lid to inspect the plate. Runs are complete when the dye tracks to the end of the trough.

(vii) Immediately transfer the gel to a shallow damp storage box on the level bench (board) and fill the troughs with about 150 μl of the appropriate antisera and cover. Do not move the box (i.e. to refrigerator) until the serum has entered the gel.

(viii) Precipitation reactions are normally left to develop overnight at 4°C. If left at room temperature examine after a few hours and stop the reaction by washing before the arcs become fuzzy. Plates are best photographed in the wet condition with diffuse underlight and the troughs filled with water. The underside of the glass can be wiped with glycerol to reduce smearing. Plates can be washed, dried and stained following the protocol for gel diffusion (Section 3.1.4). Extensive washing is needed to remove serum proteins from the gel. Background protein staining is reduced by using the IgG fraction of serum as the source of antibody.

### 4.1.4 *Technical notes*

*(i) Running and gel buffers.* Most proteins migrate anodally at pH 8.6 and for this reason a 0.05 M barbitone buffer, pH 8.6, is used routinely for serum and other protein analysis with the well in the cathode side of the plate. However, buffers such as borate or sodium acetate of lower pH, down to 4, can be employed to achieve separations of molecules with other pI characteristics without prejudice to the precipitation step or gel integrity. Optional buffer conditions should be determined by trial. Buffers of low molarity may

cause pH fluctuations and gels may overheat in buffers of too high molarity. Buffers of 0.02 to 0.075 M can be used successfully. The choice of buffer anions has been shown often to be more important than pH to the resolution of complex antigen mixtures. However, it is also apparent that achieving a separation of proteins clustered at, for example, pH 8.6 is best effected by altering the pH to be close to the proteins' pI values, to maximize the effect of minor charge differences (12). Molecules with poor solubility in water (e.g. from cell membrane extracts) can be successfully electrophoresed if the running and gel buffer includes suitable solvents such as SDS.

(ii) *Properties of agar and agarose and electroendosmosis.* At alkaline buffer pH, an agar matrix carries net negative charge through the ionization of inorganic acid groups. This produces a fixed matrix charge which is balanced by the buffer solvent $H_3O^-$. The influence of this is that the gel fluid phase and its dissolved molecules have a net cathodic flow referred to as electro-osmosis (12) or electroendosmosis. Commercial brands of agar, as opposed to agarose, have in this manner high electroendosmosis properties and as a result the least negatively charged (lowest pI molecules) in serum at pH 8.6, the $\beta$2 and $\gamma$ globulins, migrate towards the cathode (*Figure 11*). This may be no disadvantage if, by choice of buffer anions, the antigens under study can be adequately separated at both cathodic and anodic regions of the gel. For agar gels it is preferable to cut the wells in the centre of the plate to allow for cathodic protein migration. Resolution of immunoglobulin classes is often best achieved with agar gels.

In contrast to agar, some commercial grades of agarose for electrophoresis are prepared (refined) to provide smaller overall matrix charge and have low electro-endosmosis (LEE) properties. In consequence LEE agarose gives minimal migration of $\gamma$ globulins at pH 8.6, but good resolution of anodal antigens. LEE agarose is ideal for methods in which antigens are electrophoretically migrated into static antibody gels at pH 8.6 (RIEP and 2D-IEP, Sections 4.2 and 4.3), or gels where the antibody migrates towards strong cathodally migrating antigen (counter-current IEP, Section 4.4).

(iii) *Reagents.* Wells are filled with 2–5 $\mu$l of antigen. Serum can be used undiluted (for low concentration antigen detection) or diluted from 1:2 to 1:5. For single proteins, 5 $\mu$l of 1 mg/ml solution usually gives good results with strong antisera.

The amount of antiserum applied to troughs should be adjusted to give sharp precipitation arcs contained within the gel. The pattern of developed arcs will depend on the balance of antigen and antibody achieved in the gel. A good undiluted antiserum will achieve this using 150 $\mu$l per trough. In some cases it may be necessary to refill troughs after an hour or more—the alternative is to reduce the antigen loading if this still allows the full expression of precipitation arcs. Whole serum, especially if it contains haemoglobin or has high lipid content, may make IEP plates difficult to wash and background protein staining may obscure the results. IgG preparations of sera are preferable for this reason.

(iv) *Plate design.* In assessing antigens or antisera, reference reactions with known specific antibodies and standard antigens should be incorporated in the analysis. Thus, in looking at the purity of a serum protein fraction by IEP, a reference whole serum/anti-whole serum reaction should be performed at the top of the plate and tests intercalated with a known specific reaction to the standard antigen. Contaminants can be identified often by position and appropriate specific sera used to confirm the rogue molecules.

Likewise antisera under test should be compared with a certified specific reagent.
(v) *Buffers*. Buffer reservoirs should be changed regularly for fresh solution and the tanks thoroughly cleaned. Old buffer will contain contaminants which can move into gels. Likewise wicks are a source of contamination and should be changed regularly. In frequent use of IEP boxes, the polarity of the runs should be reversed on each occasion to keep the buffer in balance. Remember to change the orientation of the plates to coincide. The efficiency of buffer can be checked by measuring voltage across the plate using electrode probes.

### 4.1.5 *Advantages and disadvantages of IEP*

As a qualitative approach for analysis of mixed antigens such as serum, IEP is recognized as the definitive test system capable of discriminating at least 20 components in the gel. Individual proteins can be studied for abnormalities as the separation step reveals altered electrophoretic properties (e.g. homogeneous pI of myeloma proteins), gross changes in concentration (intensity of precipitin arcs) and subunit composition changes (distortion of $\kappa$:$\lambda$ light chain ratios in myeloma proteins). Such features are not readily detectable by simple gel diffusion assays. By prudent selection of agar/agarose matrix, buffer and pH conditions, molecules of widely differing charge can be studied. As with all precipitation assays the value of the test is restricted to molecules existing at less than or equal to 5 $\mu$g/ml in solution, and to molecules of precipitable size.

## 4.2 **Rocket immunoelectrophoresis**

RIEP is a development of the combined electrophoresis and immunoprecipitation principle, that allows a rapid estimation of single antigen concentrations by reaction with unispecific antisera.

### 4.2.1 *Principle*

Individual antigens are electrophoresed from wells at the base of a gel plate into a specific antibody-containing agarose under conditions where the antigen migrates towards the anode and the antibody (IgG) is immobile. Sharply-pointed precipitation peaks (rockets) form as the antigen accumulates bound antibody during migration. The height of the rockets is proportional to the antigen concentration and when dilutions of standard antigen are used on part of the plate, antigen measurements in unknown (test) samples in other wells can be determined by reference to a standard curve. A unispecific antibody is required in the gel for measuring single antigens in mixtures. Use of a polyspecific antiserum can test the purity of migrated antigen. The unispecificity of an antiserum can be confirmed by running against antigen mixtures. The test is restricted to antigens that migrate anodally at pH 8.6 but for these the system allows rapid (1–2 h) quantitations.

### 4.2.2 *Method*

(i) Prepare an antibody agarose plate as for RID (Section 3.2) although the antibody concentration may differ and needs to be determined by trial. Antiserum at 1–2%

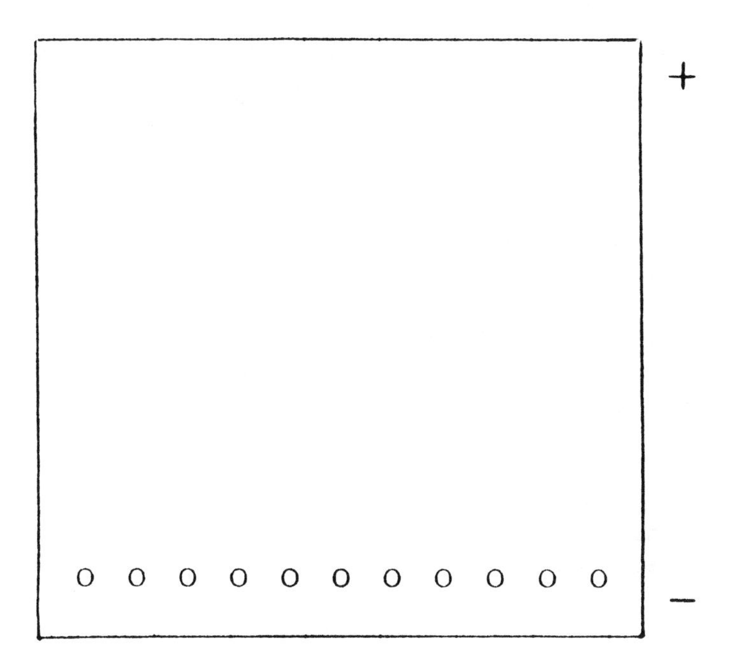


**Figure 13.** Actual size template for performing RIEP on 8 cm × 8 cm glass plates.

(v/v) is the usual range. The IgG fraction of antiserum is much preferable to whole serum as this reduces background protein staining.

(ii) When the gel is set (preferably left overnight at 4°C for antibody equilibration) place the plate on the template (*Figure 13*) and cut a row of 11 wells using the 5 μl punch. Remove the gel plugs.

(iii) Prepare a range of standard antigen dilutions—a series of five from 1 mg/ml to 10 μg/ml is recommended for accuracy. Prepare the three unknown samples (per plate) at two dilutions, 8-fold apart. Randomize the positions of standards and test samples and apply 5 μl of each to the wells.

(iv) *Immediately* electrophorese, with the anode at the top of the plate. Water cooling is essential for plates run fast at high voltage. When run at 80–100 V (10–12 V/cm) the run should be completed in 1–2 h. To ensure completion observe the leading tip of the highest standard. The run can be stopped when this develops to a sharp point (*Figure 14*).

4.2.3 *Interpretation*

Immediate running of plates is essential to prevent lateral diffusion of the antigen. Rapidly run plates should give rockets with widths no greater than the well diameter and this leads to greater accuracy. Standards should be selected that give rockets that run to completion over the dimension of the plate and with the top standard using the length of the gel. Double or multiple rockets indicate that the antiserum is not unispecific (see Figure 3, Chapter 2).

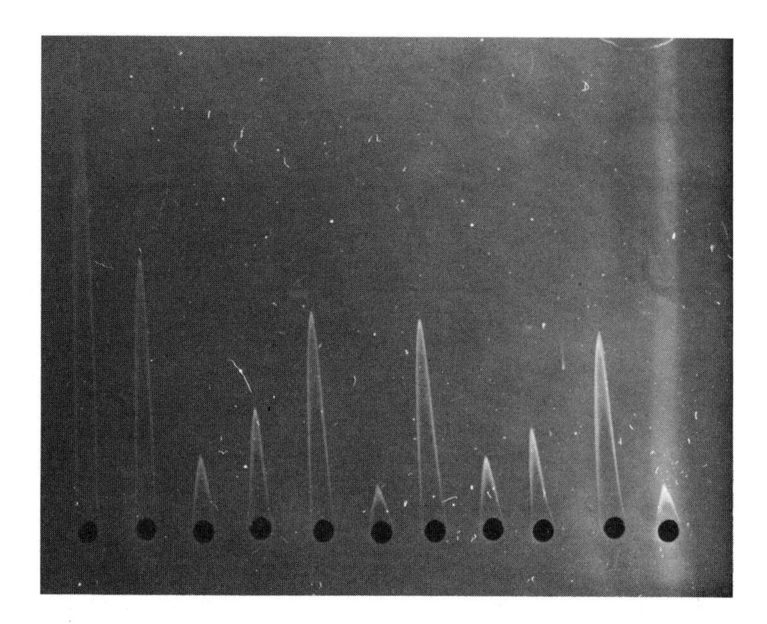

**Figure 14.** Example of a developed RIEP plate with five standard antigen dilutions and three unknown sera at two dilutions. The test is measuring human serum albumin. The agarose gel (9.8 ml) contains 80 $\mu$l of the IgG fraction of sheep antiserum to albumin. The standard antigen series is from 500 $\mu$g/ml in doubling dilutions to 31.25 $\mu$g/ml. The test sera are diluted 1:250 and 1:500.

#### 4.2.4 *Quantitation*

(i) For immediate determinations, rocket heights can be measured by placing the gel on a horizontal glass plate with light beneath. The rockets should be measured from the tip to the leading edge of the well. Alternatively gels can be washed extensively in saline for 2–3 days and then in distilled water, dried and stained before measurement.

(ii) For quantitation of antigen in unknown samples, a standard curve is constructed of peak height in increments of 0.5 mm against standard antigen concentration. Unknowns are measured against the curve. Sensitivity is in the range of 5 $\mu$g/ml. This can be improved by reducing antibody concentration and lowering the standards accordingly. In this case plates may require staining to achieve visibility of rockets for accurate measurement. Precision is in the range of 5–10%.

#### 4.2.5 *Advantages and disadvantages*

The major advantage of this method of antigen quantitation is the rapidity of results obtained, compared with RID. However, the assay is restricted to molecules of moderate to strong negative charge at pH 8.6 and this excludes immunoglobulins unless special electrolytes and agar are employed. It is possible to increase the negative charge of immunoglobulins as test antigens by carbamylation of both test sera and standards. This is achieved as follows (13).

(i) Add 1 vol of serum to 2 vol of 2 M KCNO.

(ii) Stand at room temperature for 6–18 h.

In practice it is better to estimate immunoglobulins and complement components by RID. Where rapid results are needed, or a large throughput of samples is necessary, nephelometry and related turbidometric techniques should be used (14,15).

### 4.3 **Two-dimensional immunoelectrophoresis**

#### 4.3.1 *Principle and applications*

This method combines the electrophoretic separation of antigens as a first step, in one dimension, with the rocket principle as a second step in the second dimension (at 90°). Separated antigens migrate into multispecific antibody–agarose to produce a pattern of rockets. The method is of broad analytical value in examining the complexity of antigens with quantitative features. It is valuable for testing anodally-migrating antigen purity and the specificity of antibodies. As all proteins entering the antibody gel that do not precipitate can be electrophoresed out of the plate on the anode side the method can be used to purify antigen–antibody complexes as precipitin peaks which can be used for immunization. This gives access to a very large number of antigens which are difficult to purify by other means. Comparison of antigen performance on antibody-standardized plates allows multiple antigen quantitation from which, for example, disease-related changes in serum can be determined. Changes in individual proteins, such as the activation (enzyme cleavage) of complement C3 component, and proteolysis of other proteins, can be monitored. The method can be used to test the specificity of antisera incorporated in the gel of the second dimension.

#### 4.3.2 *Method*

(i) *1st dimension.*

(1) Prepare an agarose–barbitone buffer plate as for IEP and GDD (Section 3.1.3). Do not use Gelbond.

(2) Using the template (*Figure 15*) cut the gel into 1 cm or 1.5 cm strips and cut wells in the position indicated.

(3) Fill the wells with antigen and electrophorese as for IEP (Section 4.1.3), stopping the run when the Bromophenol Blue tracks to about 1 cm from the anodal edge.

(4) *Immediately* transfer the antigen strips to the edge of a warmed new plate (this may have Gelbond applied) and pour onto the plate 8.5 ml of antibody–agarose mixture at 56°C (already prepared). The new gel should fuse evenly with the first dimension gel strip. This step should be performed as rapidly as possible to reduce the diffusion of separated antigen zones of the transferred agarose strip. A secotome blade is ideal for transferring agarose strips to new plates.

(ii) *2nd dimension.*

(1) Immediately the fused gel plate has set replace it on the cooling plate of the electrophoresis box, with the first dimension strip connected to the cathodal buffer wick (90° to first dimension run). For the second dimension electrophoresis a current of 20 V (2.5 V/cm) is recommended for overnight running (16 h) of the gel. Where antigens are to be used for immunization it is necessary to run plates for 24–48 h.

(2) Wash the gels extensively in saline and distilled water, press-dry them and stain them (Section 3.1.4). For preparation of antigens for immunization, place the

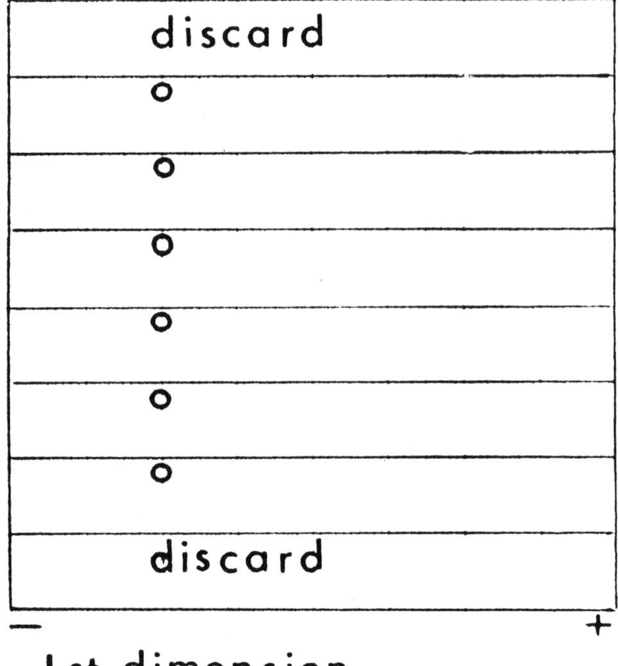


**Figure 15.** Actual size template for the first dimension electrophoresis separation of antigens to perform 2D-IEP using 8 cm × 8 cm glass plates.

washed gel over a light box and excise the selected precipitations and soak for 1–2 weeks in changes of saline.

Examples of 2D-IEP plates are shown in Chapter 2 (Figures 2 and 7).

4.3.3 *Technical notes*

(i) Reagents. 2 μl wells are normally used for the first dimension. For serum, 1 μl (undiluted) is the usual volume. 200–500 μl of antibody is the usual range added to agarose for the second dimension (2.5–6.0%), but this depends on the balance of the antibodies and the antigen concentration. Background protein staining of plates is greatly reduced by using the IgG fraction of the antiserum.

(ii) It is very important to minimize the time between running the first and second dimension. Use of a cooling plate for the first dimension is essential to achieve concentrated antigen zones by rapid separation. Cooling in the second dimension is advantageous.

(iii) Completion of the second dimension run is achieved when the highest precipitation peaks have come to sharp points. In serum runs albumin is the highest peak and the most anodal (except for pre-albumin). Good first dimension separation can be monitored by position of the albumin peak. An adequate volume of anti-whole serum protein antibody is demonstrated by the albumin peaking at the top of the plate.

(iv) For analysis of serum protein abnormalities it is essential to standardize the protocol and to use very high quality, balanced multispecific antiserum.

(v) If the technique is being used to prepare antigens, scrupulous cleanliness must be exercised at all stages. The buffer tanks must not contain extraneous proteins (run off previous gels), bacterial contaminations, etc, and fresh lint wicks must be used for each plate.

(vi) When testing antigen purity or antibody specificity, ten times the normal concentration of antigen and antibody should be used.

#### 4.3.4 *Advantages and disadvantages*

The method reveals more antigens than IEP with the same antisera and allows antigen purification as an additional facility. It is uniquely valuable as a method for testing antibody specificity with antigens undenatured by electrophoresis. By comparison sodium dodecyl sulphate (SDS)−PAGE analysis with Western blot immunostaining, although superior in sensitivity, reveals only antigens with epitopes intact after denaturation.

### 4.4 **Counter-current immunoelectrophoresis (C-CIEP)**

#### 4.4.1 *Principle and applications*

Antigens with higher pI values (negative charge) than antibody at pH 8.6, have sufficient anodic mobility on electrophoresis to migrate towards antibody moving cathodally from an opposing well in a gel with high electroendosmosis properties. Exploitation of this phenomenon allows rapid formation of a precipitation line between the antigen and antibody wells where proportions of the meeting reagents is appropriate. A very sensitive and rapid assay for both antigen and antibody is the result (16). The method has broad applications in testing for antigens of infections in body fluids (e.g. hepatitis B antigens), unusual levels of serum proteins such as $\alpha$-feto protein, and for antibodies to a range of anodic antigens, with a sensitivity of less than 400 ng/ml.

#### 4.4.2 *Method*

(i) Prepare gel plates in the standard way using a choice of agars. Difco Special Agar-Noble (1.2%) is recommended. This gives cathodic migration properties to antibodies in barbitone buffer, pH 8.6.

(ii) Cut wells using a template (*Figure 16*), usually in two columns of 2 mm wells, 6 mm apart (inner dimension). Fill the wells with antigen (standard or test) in one column (cathode side) and antibody (reference or test serum) in the other column (5 $\mu$l per well) and set up the plate for electrophoresis with the antigens running towards the anode. Different dilutions of antibody and antigen should be arranged between the opposing wells to ensure that one pair has optimal conditions for precipitation. Different antigen well sizes and volumes can be employed in the search for ideal conditions.

(iii) Perform electrophoresis with constant current at 1.5 mA/cm. Reactions should be developed after 30 min.

(iv) Plates can be washed, dried and stained (Section 3.1.4) to increase the sensitivity of reading.

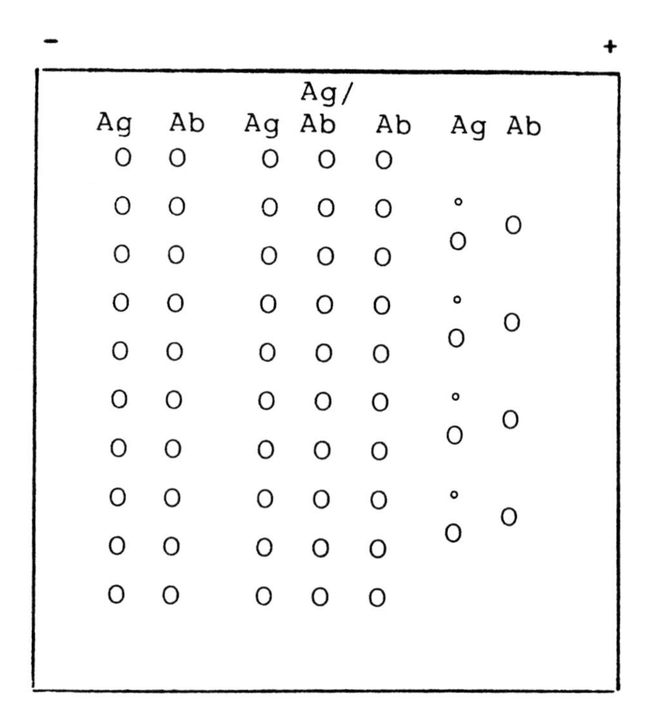


**Figure 16.** Actual size template for performing counter-current immunoelectrophoresis (C-CIEP) on 8 cm × 8 cm glass plates, using an agar and buffer system in which antibodies migrate towards the cathode and antigens have anodic migration. The two columns of wells on the left can be used to test for either antigen or antibody according to whether standard antigen (antibody test) or reference antibody (antigen test) are used. The triple column system in the centre can be used to test simultaneously for the presence of both antigen and antibody in test serum samples, where standard antigen is in the left column and reference antibody in the right, the test serum in the centre wells. The patterns on the right can be used for testing antigens which may be present at widely varying concentration. Very high concentration can be detected in proportions optimal for between-well precipitation if small volumes (1–2 μl) are used in the small wells.

## 5. WESTERN BLOTS AND IMMUNOSTAINING OF ANTIGEN BANDS

### 5.1 **Principle and applications**

Molecules separated by electrophoresis in polyacrylamide gel can be moved by transverse electrophoresis from the gel onto a nitrocellulose membrane to which they bind in identical pattern. This process is referred to as Western blotting or electroblotting (10,17). After blotting, reference antigen and molecular weight marker tracks can be stained for protein or carbohydrate and the other tracks exposed to antibody probes (9,18). Discrete antigen bands (antigenic subunits of complex molecules in reducing gels) are revealed by use of labelled antibodies. This second step is referred to as immunostaining. Overall the method exploits both the superior capacity of PAGE for separation of molecules in low concentration and the ability of antibodies to reveal specific antigenic components. The method thus has major applications in the identification of antigenic molecules by molecular size and composition but is also an invaluable approach to testing the specificity of antibodies [particularly monoclonal antibodies (Mabs)] and in search for antigenically related molecules in the same and different antigen preparations (e.g. between different bacterial species, viral mutants, cell extracts, etc.).

SAMPLE ON STRIPS:BCG.son. SAMPLE ON STRIPS:MTB.son.

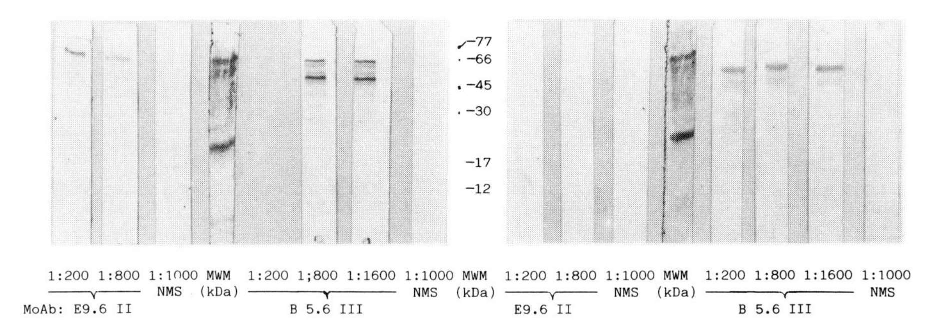


SAMPLE ON STRIPS : BCG.son. SAMPLE ON STRIPS : MTB.son.

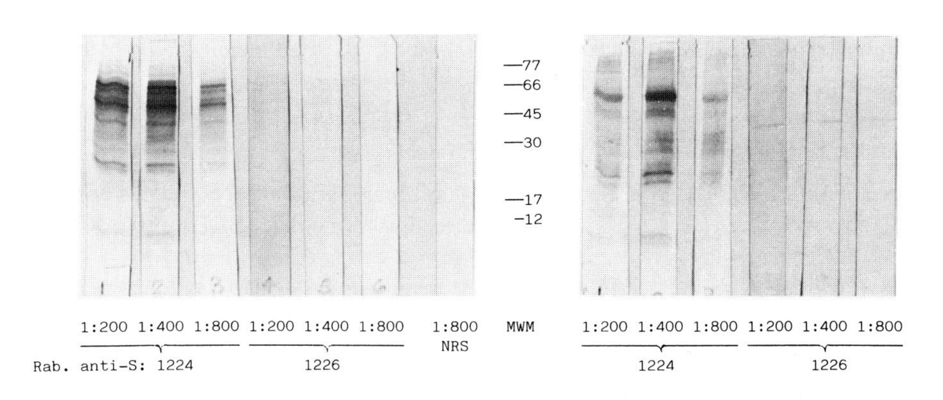


RABBIT ANTI-SERA: 1224( from rabbit immunized with sonicated M.tuberculosis )
1226( from rabbit immunized with sonicated M.bovis (BCG.strain))

**Figure 17.** Examples of PAGE Western blot immunostaining. Replicate tracks of sonication extracts of *M.bovis* BCG (**left side, upper** and **lower** plates, BCG) and *M.tuberculosis* (**right**, MTB) have been separated in a reducing 14% polyacrylamide gel, blotted onto nitrocellulose membrane and the tracks, including the mol. wt markers (MWM; kDa), cut into separate strips and individually immunostained with antibodies (the MWM tracks stained for protein). The upper set are stained with dilutions of two monoclonal antibodies (Mab), normal mouse serum (NMS) being used as control. These strips were developed with HRP-conjugated anti-mouse immunoglobulin and diaminobenzidine substrate. The lower set are stained with two rabbit antisera, as indicated, with normal rabbit serum (NRS) as a control. Development was with HRP-anti-rabbit IgG.

## 5.2 Equipment and reagents (see Appendix)

Electroblot apparatus with cooling system.
Nitrocellulose membrane.
Heavy grade filter paper.
Thin polythene sheet (e.g. 'Clingfilm').
Washing and staining trays.
Razor, scalpel or cutter to prepare membrane strips.
Shaker incubator.
Latex gloves.

Dish for immersing polyacrylamide gel.

*Blotting buffer*

| | | |
|---|---|---|
| Glycine | 2.93 g | (39 mM) |
| Tris | 5.81 g | (48 mM) |
| SDS | 0.375 g | (0.0375% w/v) |
| Methanol | 200 ml | (20% v/v) |

Add distilled water to 1 litre.

*Washing buffer*

PBS–Tween; PBS, pH 7.2 (Section 2.3) containing 0.05% (v/v) Tween 20.

*Blocking buffer*

PBS–Tween containing 5% (w/v) dissolved non-fat milk powder.

*Primary antibodies*, e.g. murine Mabs as ascitic fluid or culture supernatant, polyclonal antisera (e.g. rabbit, sheep, etc.).

*Secondary antibody (antiglobulin)*, for example HRP-conjugated sheep anti-mouse immunoglobulin, HRP-conjugated sheep anti-rabbit IgG, HRP-conjugated rabbit anti-sheep IgG.

*Enzyme substrates*

Useful substrates with insoluble coloured products are:

(i) 3-Amino/4-ethylcarbazole (AEC)—reddish pink. AEC buffer is 0.05 M acetate, pH 4.5. Dissolve 4 mg in 200 $\mu$l of dimethyl formamide (BDH) plus 10 ml of buffer and filter.
(ii) Diaminobenzidine (DAB)—red/brown. Filter before use. DAB buffer is 50 mM Tris, pH 7.4. Use 0.5 mg in buffer.

Both enzyme substrate systems require $H_2O_2$ (5 $\mu$l/20 ml).

*Autoradiography*

Radio-iodinated affinity-purified antiglobulin reagent (see Volume II).
X-Ray films and film cassette.
Intensifier screen (optional).
Film developer (Kodak D19) and fixer.

## 5.3 **Method**

### 5.3.1 *Electroblotting step*

Carry out all procedures using latex gloves to avoid fingerprints on the nitrocellulose membrane.

(i) Separate the electrophoresed polyacrylamide gel from the supporting glass plate in a dish of distilled water.
(ii) Transfer the gel onto thick filter paper soaked in blotting buffer and place this on top of several more layers of soaked filter paper.
(iii) Cut a square of nitrocellulose membrane of appropriate size; soak in blotting buffer and place on the surface of the gel. Apply several layers of soaked filter paper above the membrane. Use a test tube as a roller to remove all air and establish close contact throughout the layers.

(iv) Place the electrode plates on each side of the assembled sandwich, ensuring that the anode plate is on the nitrocellulose membrane side; fasten the assembly together according to the kit instructions and place in the electrophoresis unit so that transverse current moves the molecules from the gel onto the membrane by anodic migration. Turn on the cooling system (or remove the unit to the cold room).

(v) Turn on the current. The required field strength is determined by the surface area and thickness of the gel—0.8 mA per $cm^2$ is a useful guide. The recommended time for electrophoresis of a 1.0 mm gel is 45–60 min with a modern apparatus such as the LKB Multiphor II. Longer periods up to 16–24 h may be necessary with less efficient, non-cooling apparatus. The Multiphor system allows several gels to be blotted simultaneously. In this case each set of gel plus membrane is separated by a film of polythene (e.g. Clingfilm) soaked in buffer.

(vi) After blotting, the nitrocellulose membranes are stored in washing buffer (PBS–Tween) between buffer-soaked filter paper prior to staining. The presence of Tween 20 is essential to block non-specific binding of immunoglobulins in the immunostaining step.

5.3.2 *Immunostaining step*

There are two development systems for immunostaining, one utilizes radio-iodinated antibody followed by autoradiography, the second utilizes the enzyme-linked immunosorbent assay (ELISA) principle of enzyme-conjugated antiglobulin with a localized colour reaction due to deposition of insoluble coloured substrate product onto the antigen/antibody bands on the nitrocellulose (10). Problems of non-specific binding are equivalent in the two staining systems and are countered in the same manner. The enzyme method, because of its simplicity and stable reagents, is preferred except where extreme sensitivity is required. The enzyme method is described below but reference to the autoradiographic alternative is made in the technical notes. Methods for preparing antiglobulin reagents are provided in Chapter 2. Conjugation of antibody to enzymes (e.g. horseradish peroxidase, HRP, is described in Volume II). HRP-Conjugated antiglobulin reagents to a wide variety of species antibodies are available commercially (see Appendix). An example of the developed enzyme-staining method for Western blots is shown in *Figure 17.*

(i) Wash the Western blot membrane for at least 1 h in PBS–Tween and then cut the tracks into strips and number them for orientation. Stain the standard antigen and molecular weight marker strips for protein (or carbohydrate) as described in Chapter 2 [Section 3.2.2 (vii)]. These confirm that efficient blotting has occurred and serve as the markers for the antibody reactions.

(ii) Place the strips for antibody staining in small trays and submerge them in the primary antibody diluted in PBS–Tween. Staining proceeds for 30–60 min at 37°C, using preferably a rocking incubator. For standardization, it is advisable to adopt a uniform washing and staining time. 1 h is recommended.

(iii) Wash the strips thoroughly for 30–60 min in PBS–Tween and then immerse in HRP–antiglobulin conjugate diluted in blocking buffer. Staining proceeds for a further 30–60 min at 37°C with rocking.

(iv) Wash the strips again and then immerse in substrate solution in the appropriate buffer, in small dishes in the shaker where the staining reaction is to be observed. Exposure to bright light must be avoided at this stage.

(v) On development of bands halt the reaction by washing the strips in PBS–Tween. Dry the strips on filter paper and assemble against the reference protein-stained tracks. They can then be conveniently glued to a mounting card for preservation.

5.3.3 *Technical notes*

(i) The dilution of HRP–antiglobulin must be determined by trial. In our hands the conjugate is the major source of non-specific staining and this is controlled by using the maximum dilution that gives a good intensity of positive staining and by use of the blocking reagent (fat-free mik) at this stage followed by extensive washing, as traces of milk left on the membrane block the substrate. It should also be noted that in looking at antigens of microorganisms the antiglobulin serum may have cross-reacting antibodies to the antigens separated on the gel. This unwanted specific binding may need to be removed by absorption of the conjugate. Generally a dilution of conjugate less than half that used in ELISA plates to give an optical density reading of 1.0 in 20–30 min, when reacted against 1 μg/ml of coating target immunoglobulin, is suitable for Western blot immunostaining. This will usually be in excess of 1:1000 dilution, where for ELISA the conjugate would be used at 1:5000.

(ii) The specificity of the antiglobulin conjugate needs to be carefully considered when used to demonstrate Mab binding to antigen blots. Each monoclonal has a restricted isotype. It is possible to use an anti-mouse immunoglobulin reagent that reacts well with all mouse isotypes but this should be confirmed by ELISA using purified mouse isotypes to coat ELISA wells, or by testing a set of Mabs of all isotypes in standard Western blot assays.

(iii) The dilution of the primary antibody used to bind the blotted antigens may be critical. At too high a concentration there may be unacceptable degrees of non-specific binding of immunoglobulin to protein bands on the blot which makes interpretation difficult. It is advisable to run several tracks of the same antigen on the PAGE run and test the antibody at several dilutions. Positive staining is remarkably sensitive and bands can be stained with as little as 1 μg/ml antibody. For monoclonal ascites this may be a dilution greater than 1:1000.

(iv) Do not expect all Mabs with proven activity against antigen by other tests to give positive Western blot immunostaining. The disruption of antigen molecules under conditions of PAGE is such as to destroy many epitopes. Estimates vary between 10% and 30% of Mabs to complex protein antigens that stain at any dilution in this system. Conversely some antibodies appear to stain well on blots of denatured antigens. *Figure 17* illustrates the pattern of staining of some Mabs to myobacterial antigens of the bacterial cell wall which have retained antigenicity under the denaturing and reducing conditions of the PAGE run.

(v) The autoradiographic technique can utilize radio-iodinated affinity-purified antiglobulin reagent or labelled *Staphylococcus* Protein A. In the latter case, it should be recognized that only some antibody isotypes bind to Protein A (19). With this reagent and antiglobulin, approximately $10^5$ c.p.m./ml should be applied to the membrane strips. After staining for 30–60 min, the strips should be washed very extensively with repeated changes of PBS before the autoradiographic step. Methods for radiolabelling of antibodies are considered in Volume II.

(vi) The autoradiographic step is performed as follows:

(1) Under safety light illumination (Kodak 6B dark brown) in a darkroom, the antibody-labelled nitrocellulose strips are arranged on a glass plate and overlaid with a sheet of X-ray film and a second glass plate. The assembly is taped together and placed in a light-proof metal cassette or sealed in a black polythene bag which is secured under a lead shield. A range of X-ray films are suitable (see Appendix). Films with emulsion on both sides are required for use with an intensifying screen; this allows development on both sides of the film to improve sensitivity with critically small amounts of high energy emissions. The intensifying screen is placed on the other side of the film to the nitrocellulose.

(2) Film exposure may be for 1–14 days according to the isotope (i.e. $^{125}$I or $^{131}$I) and the specific activity of the antibody probe. With the intensifying screen the assembly must be kept at $-70°C$ during exposure.

(3) The film is developed under safety light with Kodak D19 developer for about 5 min at room temperature.

## 6. REFERENCES


1. Ouchterlony,Ö. (1958) In *Progress in Allergy*. Kallos,P. and Waksman,B.H. (eds), Karger, Basel, Vol. VI, p. 30.
2. Elek,S.D. (1948) *Br. Med. J.*, **1**, 493.
3. Grabar,P. and Williams,C.A. (1953) *Biochem. Biophys. Acta*, **10**, 193.
4. Laurell,C.B. (1966) *Anal. Biochem.*, **15**, 45.
5. Laurell,C.B. (1972) *Scand. J. Clin. Lab. Invest.*, **29**, 124, Suppl.
6. Ressler,N. (1960) *Clin. Chim. Acta*, **5**, 795.
7. Mancini,G., Vaerman,J.P., Carbonara,A.O. and Heremans,J.F. (1964) In *Protides. Biological Fluids*. Peeters,H. (ed.), Pergamon Press, Oxford, Vol. II, p. 370.
8. Mancini,G., Carbonara,A.O. and Heremans,J.F. (1965) *Immunochemistry*, **2**, 235.
9. Burnette,W.N. (1981) *Anal. Biochem.*, **112**, 195.
10. Gershoni,J.M. and Palade,G.E. (1983) *Anal. Biochem.*, **131**, 1.
11. Stites,D.P. and Rogers,R.P.C. (1987) In *Basic and Clinical Immunology*. Stites,D.P., Stobo,J.D. and Wells,J.V. (eds), Appleton and Lange, Norwalk/Los Altos, p. 241.
12. Williams,C.A. (1971) In *Methods in Immunology and Immunochemistry*. Williams,C.A. and Chase,M.W. (eds), Academic Press, New York, Vol. III, Chapter 14, p. 238.
13. Milford-Ward,A. (1980) In *Techniques in Clinical Immunology*. Thompson,R.A. (ed.), Academic Press, 2nd edition, p. 1.
14. Osler,A.G. (1971) In *Methods in Immunology and Immunochemistry*. Williams,C.A. and Chase,M.W. (eds), Academic Press, New York, Vol. III, Chapter 13, p. 83.
15. Nilsson,L.-Å. (1981) In *Immunoassays for the 80s*. Voller,A., Bartlett,A. and Bidwell,D. (eds), MTP Press, Lancaster, UK, Chapter 5, p. 43.
16. Bussard,A. (1959) *Biochim. Biophys. Acta*, **31**, 258.
17. Towbin,H., Strahelin,T. and Gordon,J. (1979) *Proc. Natl. Acad. Sci. USA*, **76**, 4350.
18. Haid,A. and Suissa,M. (1983) In *Methods in Enzymology*. Fleischer,S. and Fleischer,B. (eds), Academic Press, New York, Vol. 96, p. 192.
19. Johnstone,A. and Thorpe,R. (1987) *Immunochemistry in Practice*. Blackwell Scientific Publications, Oxford, 2nd edition.

CHAPTER 7

# Haemagglutination and haemolysis assays

NOEL R.LING and DAVID CATTY

## 1. INTRODUCTION

All microparticulate antigens (e.g. bacteria, yeasts, leukocytes and erythrocytes) may be agglutinated by antibodies to antigenic determinants on their surface and this phenomenon has been exploited widely in the study of cell and microorganism surface antigens and in the detection of antibodies. The use of serum complement with antibodies allows, in some circumstances, a further set of tests based on cytotoxicity and target cell lysis. The lytic reaction is best observed with red cells, and haemolysis has been used to measure both lytic antibodies and complement levels, the latter as a complement consumption test.

Erythrocytes are natural targets for antibodies in man (e.g. the isoagglutinins to the ABO system) as well as the target for antibodies in some autoimmune conditions and in materno–fetal rhesus antigen incompatibility (see Volume II). In such cases red cells are used as the source of antigen. In this chapter we provide protocols for the broader application of red cells for haemagglutination and haemolytic assays of antibodies. In principle red cells can be bound not only by antibodies specific to the red cell membrane antigens themselves (direct tests) but also to antigens either chemically bound or physically adsorbed to the cell surface. The recognized convenience and sensitivity of direct haemagglutination and haemolysis tests encouraged attempts to couple soluble antigens to erythrocytes and the success of this approach has led to the development of versatile passive test systems.

It is generally recognized that the red cell is the most suitable 'carrier particle' for passive agglutination even though the system has frequently been regarded as old-fashioned. Plastic particles made of polystyrene, polyacrylamide or aminopolystyrenes have been used successfully for the binding of proteins (1,2) but agglutination tests in which they have been used are crude and less sensitive than haemagglutination (3,4).

The size, shape, stability and agglutinability of red cells of different species varies considerably. Human red cells are larger than those of sheep and somewhat more agglutinable, whereas bovine red cells are less agglutinable than other species and show considerable variation from animal to animal (5). Red cells of all species may be rendered more agglutinable by treatment with trypsin, neuraminidase or other enzymes that strip off the surface 'fur'. Sheep red blood cells (SRBC) are the most commonly used in laboratory tests; they require no pre-treatment and the protocols described below refer to this species. The source of the sheep blood is a very important consideration. If the blood is obtained from a commercial source the red cells are usually in a poor condition on arrival and collected at random from a slaughterhouse. It is more satisfactory to

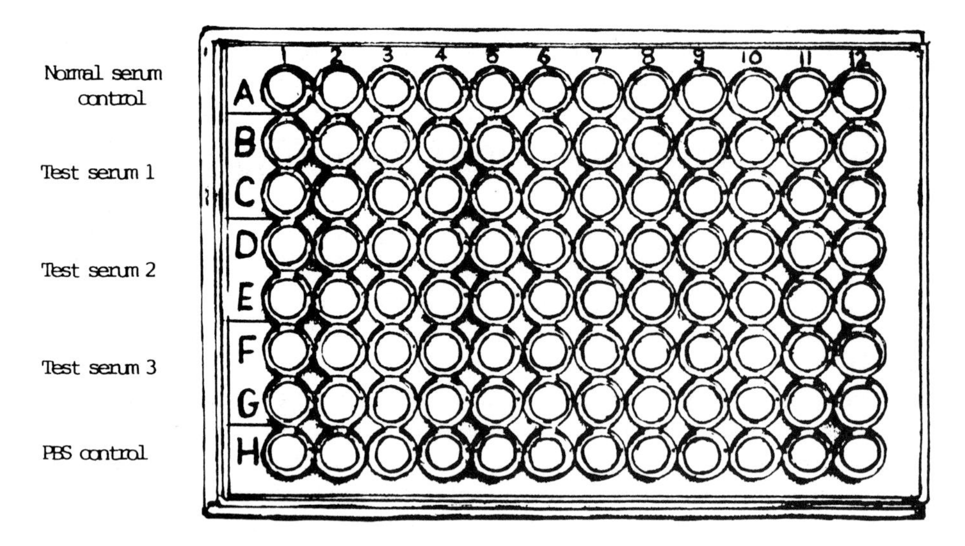


**Figure 1.** Microtitration plate with rows marked for test sera and controls for direct haemagglutination assay.

arrange for regular bleeding of a male sheep at a local centre (fragility of red cells is increased in pregnancy), selected by pre-testing blood from a number of animals.

The principle of quantitative haemagglutination assays is similar whether the test is direct or passive. Plastic microtitration plates with U-bottom wells (250 $\mu$l capacity) in a 12 $\times$ 8 pattern are normally used (*Figure 1*). Antisera are diluted down the row of wells and a suspension of red cells added. When sufficient agglutinating antibody is present to provide bridges between the cells' surface antigens the cells will clump together. As the antibody is diluted out, insufficient is available to crosslink the cells which will settle as a button on the well bottom. The end point titre of the serum is the last dilution giving a distinctive complete agglutination pattern.

In most species both IgG and IgM antibodies have agglutinating properties. Generally IgM antibodies are more efficient than IgG because of their greater binding valency. In direct and passive tests sera can be compared quantitatively for end point agglutination titre providing the conditions are exactly standardized. Some antibodies to red cell antigens in man and some that are raised in animals produce little or no agglutination in direct and passive tests. This could be the case if there are few or poorly-available antigenic determinants on the cells and if the antibody is of a form of IgG that is an inefficient agglutinin. Such antibodies, after binding to the red cell, can be detected by the use of antiglobulin sera raised to Ig determinants of the antibody-coating species. This test was developed by R.R.A.Coombs to detect so-called 'incomplete' antibodies, including the IgG rhesus specificity antibodies in man. The test is for this reason called the Coombs test and is very sensitive. It has other specialized uses—for instance in direct and passive tests it can be used to determine and measure the class and IgG subclass of antibody bound to red cells at sub-agglutinating doses, using isotype-specific antiglobulins. It is also used with standard Ig-coated red cells to compare the titre and specificity of prepared antiglobulin reagents.

2-Mercaptoethanol (2-ME) has the property of dissociating the IgM pentamer so that it can no longer produce agglutination. Addition of this reagent then leaves only the resident IgG antibodies in sera to produce agglutination. This provides a means of comparing IgM and IgG antibody titres in agglutination systems. Complement-fixing antibodies will, in the presence of active complement (fresh serum), give rise to lysis of target red cells. Whilst in most species some subclasses of IgG fix complement, IgM antibody is much more efficient and antisera with a high IgM antibody titre will have a higher haemolytic titre than sera with a predominant IgG antibody. In general the haemolysis test is more sensitive than the agglutination test if the binding antibodies are complement-fixing. As 2-ME also inhibits complement fixation by IgM, so it will greatly reduce the haemolytic titre of antisera with high IgM antibody levels and this allows the measurement of IgM antibody (see *Figure 2*).

In passive agglutination, where an extraneous antigen is attached to red cells, the system can be converted to a sensitive inhibition assay of free antigen. To do this a previously titrated agglutinating serum is used at minimum agglutinating dilution which is easily inhibited by trace amounts of test antigen in solution from agglutinating the cells. The test is quantitated by use of standard inhibitor dilutions.

## 2. DIRECT HAEMAGGLUTINATION ASSAY

A simple, sensitive method of titrating antibody to a red cell antigen is by serial double dilution of the antibody in 50 μl or 100 μl volumes in a U-bottomed microtitration plate followed by addition of a standard dilute suspension of cells (e.g. 0.5% v/v) and allowing the cells to settle at room temperature. The titre is then read by inspection of the 'settled pattern' end point. Non-agglutinated cells settle as a tight button whereas agglutination is indicated by a large circle of cells which, if agglutination is strong, may have a 'crinkly' edge. This is a very sensitive test even for low affinity antibodies provided the antigen is well expressed on the cells. The protocol below is for the titration of anti-SRBC antibodies (raised in rabbits as an example). Such antibodies are specific to several antigens on the SRBC. Although of limited application itself, the assay is an invaluable model for learning the principles of haemagglutination procedure and interpretation of results. The assay can be modified to demonstrate agglutination and haemolysis with IgG and IgM antibodies. *Figure 2* is an example of titrations to illustrate aspects of the assay system.

### 2.1 **Equipment and materials** (see Appendix for suppliers)

Microtitration U-bottomed plastic plates with covers.
Diluting tulip—50 μl (or 100 μl).
Micropipette—50 μl (or 100 μl) with disposable tips.
Bunsen burner.
Beaker of distilled water.
Filter paper.
Normal rabbit serum (pre-bleed of immunized rabbit is ideal).
Alsevers solution: dextrose, 2.05% w/v; NaCl, 0.42% w/v; trisodium citrate, 0.3% w/v; citric acid, 0.055% w/v; sterilize by filtration.
Rabbit anti-SRBC: Sample taken early in primary response for IgM antibodies. Samples taken after repeated immunization for IgG antibodies (see Chapter 2, Section 6.7.2 for preparation).

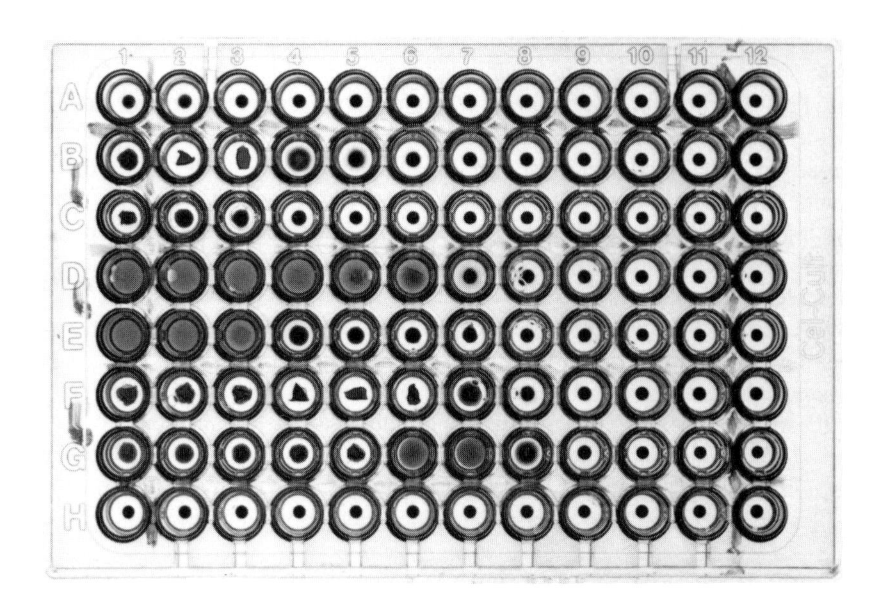


**Figure 2.** Direct haemagglutination and haemolysis assays showing the effect of pre-treatment of sera with 2-ME. The sera tested in rows B and C; D and E are from primary immunizations (mainly IgM antibody). Complement has been added to rows D and E to demonstrate the lytic activity of IgM antibody and effect of 2-ME on this (row E). The serum tested in F and G is from a secondary (IgG) response where 2-ME has less effect on agglutination (row G). Wells G6,7 and 8 are open agglutinations not lytic reactions. See Section 2.2 for design and 2.3 and 3.4.2(v) for interpretation.

Fresh SRBC stored in Alsever's solution.
Phosphate-buffered saline (PBS), pH 7.2.
0.1 M 2-ME in PBS.
Marker pen.

## 2.2 **Method**

(i) Wash the SRBC three times in PBS, removing any residual buffy coat on the first wash. Finally prepare a 0.5% (v/v) suspension sufficient for the tests required. Suspensions up to 2% can be used—these alter the pattern and end point of agglutination and haemolysis.

(ii) On a titration plate mark off the first and last rows (A and H) and then pairs of rows between (i.e. B and C; D and E; F and G) (*Figures 1* and *2*).

(iii) Pipette 50 $\mu$l (or 100 $\mu$l) of PBS into each well.

(vi) Flame the diluting tulip to incandescence and quench in the distilled water. Blot the tip onto filter paper (do not wipe) and very carefully touch the meniscus of the normal rabbit serum (do not immerse). Transfer the tulip to well 1 of row A, rotate a few times and from an upright position transfer the tulip to well 2 and so on to well 12. Finally wash the tulip and flame. Row A is the normal rabbit serum control (see *Figure 2*).

(v) Repeat with the three test sera but in this case with duplicate rows (i.e. B and C as repeats). No serum is added to row H (see *Figure 2*).

(vi) To rows A, B, D, F and H add a further volume of PBS.

(vii) To rows C, E and G add instead a volume of 0.1 M 2-ME. This reagent must be added in a fume cupboard and the plate covered from then on except when adding reagents or examining the result.

(viii) Add 50 μl (or 100 μl) of freshly suspended SRBC to each well and cover.

(ix) Hold the plate on the bench, over a white background; gently mix the cells by tapping in rotation all four corners of the plate.

(x) Allow the reaction patterns to settle on the bench for up to 2 h or overnight at 4°C. Reactions can be accelerated at 37°C although the settling of the cells is not influenced by temperature.

### 2.3 **Interpretation and technical notes**

(i) Many normal sera contain background (heterophile) anti-SRBC antibodies, usually in lower titre. Some agglutination may be seen in the first wells of normal serum dilutions.

(ii) Sensitivity of the assay is generally greater using 0.5% cell suspension than higher cell numbers.

(iii) The haemagglutination test detects IgM antibodies preferentially not only because of the multivalency of this antibody (five binding sites compared with two for IgG), but also because of the large size of IgM. As erythrocytes are further apart when agglutinated with IgM, the repulsive force due to the cells' zeta potential is less.

(iv) The effect of 2-ME treatment of sera on the destruction of IgM-dependent agglutination can be best observed in the primary anti-SRBC response (see *Figure 2*). The residual serum agglutinating properties are due to IgG antibodies. Rabbit IgG anti-SRBC antibodies are less heterogeneous in agglutinating ability than in some other species. An additional antiglobulin step may be required to demonstrate bound IgG antibody with poor or absent agglutinating potential in these species (see Section 4).

(v) A paucity of available antigenic determinants on the red cells for IgG antibodies may contribute to poor agglutination in some cases. By contrast, agglutination is particularly strong in the human ABO system because of the combined influence of the predominant IgM response and the high surface antigen density—about $5 \times 10^5$ binding sites per cell.

(vi) There should be no spontaneous agglutination of cells in serum-free control wells. This is seldom a problem with fresh cells with a high net negative charge. When erythrocytes are coupled to soluble antigens for passive agglutination assays, the surface charge is often reduced and spontaneous agglutination may occur. This problem is dealt with by using a protein-rich diluent. This is usually unnecessary with the direct test.

## 3. HAEMOLYTIC ANTIBODY ASSAY AND THE TITRATION OF COMPLEMENT

### 3.1 **Principle**

Complement is a multicomponent system present normally in serum in an inactive state. When the first component is activated by fixation to antigen–antibody complex (of complement-fixing antibody isotype), it is then able to activate molecules of the next component and a cascade fixation reaction occurs, culminating in lysis when susceptible cells such as erythrocytes are the target. When antibodies to antigens on red cells are diluted out in an excess of complement, the lytic end point is a measure of antibody activity. As complement is labile in serum on storage, accurate titrations depend on use of a quantified excess of an extrinsic source, which is usually preserved, or freshly-harvested, guinea pig serum. Surviving complement activity in the test sera is destroyed by heating as a preliminary step, to provide additional standardization. To quantify the added complement, this is titrated in a haemolytic system of (IgM) antibody-coated red cells, after absorption to remove intrinsic red cell antibodies from the guinea pig serum.

The steps in the haemolysis titration are as follows.

(i) Preparation of antibody-coated red cells for complement titration.
(ii) Test for adequate coating by lysis in an excess of complement.
(iii) Absorption of stock complement with red cells in saline in the cold.
(iv) Titration of stock complement in the lytic indicator system to define a 50% lytic dilution ($CH_{50}$), and the dilution giving 4 $CH_{50}$ units (standard excess) for use in antibody titrations.
(v) Test serum complement inactivation by heating.
(vi) Titration of test sera (e.g. anti-SRBC in direct assay) in standard complement excess.

3.2 **Materials and reagents** (additional to Section 2.1; see Appendix for suppliers)

*Haemolysis diluent (HD):*

Stock solution A: 85 g of NaCl
2.75 g of sodium barbitone
Dissolve in 1400 ml of distilled water

Stock solution B: 5.75 g of diethyl barbituric acid in 500 ml of hot distilled water

Stock solution C: 20.3 g of $MgCl_2.6H_2O$ (2 M) dissolved in 50 ml of distilled water. Add 30 ml of 1 M $CaCl_2$ solution. Adjust to 100 ml to provide 1 M $MgCl_2$ with 0.3 M $CaCl_2$

(i) Mix A and B and cool to room temperature.
(ii) Add 5 ml of C.
(iii) Adjust to 2 litres with distilled water and store at 4°C (stock diluent).
(iv) Working diluent: use 1 part stock plus 4 parts distilled water.

*Saline solution:* 0.9% (w/v) NaCl in distilled water; sterilize by autoclaving in a hard glass bottle.

*Rabbit anti-SRBC sera:*
A bleed from a primary response is required for a good lytic system for complement titrations. Samples after chronic immunization can be used to demonstrate IgG and IgM haemolytic antibody titres.

*Complement:* Preserved (commercial) or freshly-harvested guinea pig serum.

*Ammonia solution:* 0.04% in distilled water (or 10% saponin).

Water bath at 56°C.
Spectrophotometer.
Rotary mixer (in warm room or incubator at 37°C).
Ice bucket.
Glass test tubes (2 ml) and rack.
Graduated pipettes (2 ml) or variable micropipette (to 2 ml) with tips.
Universal glass bottles (20 ml).
Measuring cylinder (100 ml).

### 3.3 **Titration of guinea pig complement**

#### 3.3.1 *Principle*

Antibody-coated erythrocytes are added to dilutions of complement in test tubes and the degree of lysis achieved is measured against the haemoglobin of completely-lysed cells by spectrophotometry. The lysis approaches 100% in a sigmoid curve (asymptotically) so the lytic unit of complement for total lysis is difficult to determine. For this reason a 50% lysis unit is defined ($CH_{50}$) under standardized conditions. The conditions which influence lysis are the red cell and antibody-coating concentrations, the buffering conditions and temperature. The definition of the $CH_{50}$ unit is arbitrary, relating to the particular conditions of the test.

#### 3.3.2 *Preparing antibody-coated SRBC*

(i) Inactivate the complement of a high IgM anti-SRBC rabbit serum (1 ml) by heating at 56°C in the water bath for 30 min.
(ii) Titrate the serum by agglutination against a 1% (v/v) suspension of SRBC (Section 2) in HD.
(iii) Calculate the volume of this antiserum required to coat 10 ml of 6% (v/v) red cells at the first sub-agglutinating dilution, and the volumes required for one dilution above and one below this dose. Add the 3 calculated volumes to three Universals of 10 ml of 6% cells in HD (washed thoroughly in HD), suspend and place on a rotary mixer (or frequently suspend) for 30 min at 37°C. Store at 4°C and use within 24 h.

#### 3.3.3 *Testing for adequate antibody coating (see Figure 3)*

(i) Bring 1 ml of complement to 0°C on ice, add 2 ml of ice-cold saline and 0.3 ml of packed red cells washed in cold saline. Mix and stand at 0°C for 1 h to absorb out anti-SRBC antibodies under non-lytic conditions. Finally, centrifuge the red cells at 200 *g* for 10 min at 4°C and take off the absorbed complement (now at 1:3 dilution in saline).
(ii) Dilute a small aliquot of the three stock coated cells to 1% (v/v) in HD.
(iii) To three pairs of rows of a titration plate add 50 μl of HD and in each pair titrate out the absorbed complement with the 50 μl tulip. Leave the last three wells of the second row of each pair as controls with 50 μl of HD added instead. To each pair of rows now add 50 μl of one of the antibody-coated 1% red cell suspensions, mix the wells, cover and incubate at 37°C for 1–2 h. An appropriate

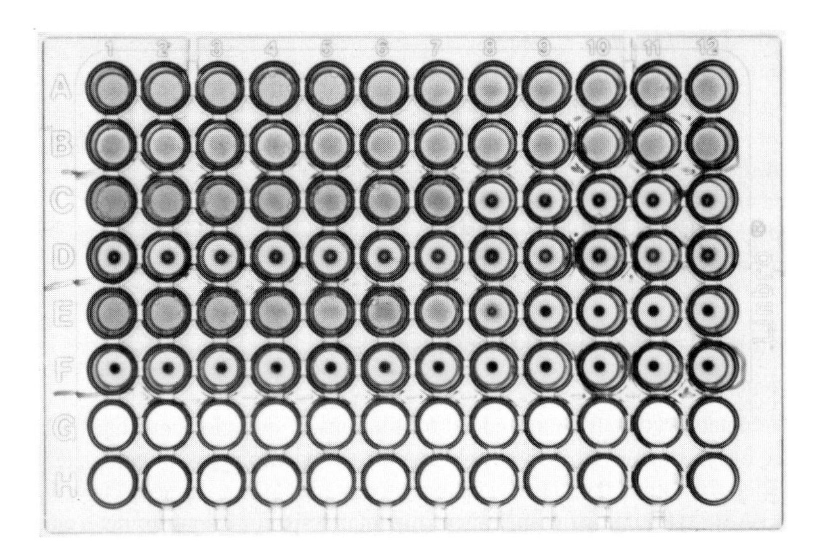

**Figure 3.** Determination of Minimum Antibody Haemolytic Dose by titration of complement with red cells coated with three concentrations of haemolytic antibody (double dilution series) in successive pairs of rows, **A** and **B**; **C** and **D**; **E** and **F**. Complement is at 1:3 in column 1 and doubly-diluted across the plate in the first row of each pair. The second rows are HD controls to test for absence of spontaneous lysis or agglutination. Cells over coated with antibody (**rows A** and **B**) spontaneously lyse on incubation. Half as much antibody (**rows C** and **D**) allows complement-mediated lysis to a complement titre of 1:192 but with evidence of spontaneous (auto) agglutination of the sensitized cells. Reduction of antibody-coating by a further dilution step gives a sensitive haemolytic system without auto-lysis or auto-agglutination. This amount of antibody can then be used to titrate complement samples and to prepare lytic indicator cells for complement fixation tests (Section 8).

coating concentration of antibody will show an extended run of lysis along the row and this defines the minimum antibody haemolytic dose (MAHD). Coating concentrations of antibody and the starting dilution of absorbed complement may need to be adjusted to achieve this result. Coating with too much antibody may result in complement-independent (spontaneous) lysis of the cells, as seen in *Figure 3*.

### 3.3.4 *Titration of absorbed stock complement: the $CH_{50}$ end point*

(i) A set of absorbed complement dilutions is mixed with a standard suspension of red cells coated with 1 MAHD in test tubes as in *Table 1*. The starting dilution of complement (in HD) should be about 1:30 [see Section 3.3.5(iii)].

(ii) Cover the tubes and incubate at 37°C for 2 h, with occasional mixing, and then place on ice for 30 min. Add 1 ml of ice-cold HD to each tube and mix. Centrifuge at 200 *g* for 10 min and then read the optical densities at 413 nm.

(iii) Plot the percentage lysis, using control tube 9 as 100%, against volume of added complement. This should yield a sigmoid curve. Determine from the curve the volume of complement needed to produce 50% lysis ($CH_{50}$). From this result

**Table 1.** Complement titration.

| | *Tube number* | | | | | | | | |
|---|---|---|---|---|---|---|---|---|---|
| | *1* | *2* | *3* | *4* | *5* | *6* | *7* | *8* | *9 (control)* |
| HD (ml) | 1.20 | 1.15 | 1.10 | 1.05 | 1.00 | 0.90 | 0.80 | 0.70 | 1.20 (ammonia) |
| Complement (1:30/ml) | 0.00 | 0.05 | 0.10 | 0.15 | 0.20 | 0.30 | 0.40 | 0.50 | 0.00 |
| 1% SRBC 1 MAHD (ml) | 0.30 | 0.30 | 0.30 | 0.30 | 0.30 | 0.30 | 0.30 | 0.30 | 0.30 |
| Lysis expected | − | + | + | ++ | ++ | ++ | ++ | ++ | ++ (100%) |
| | | − | | | + | + | ++ | ++ | ++ |

calculate the dilution of complement required for a $CH_{50}$ unit in 50 μl for an antibody titration assay using 50 μl of 1% red cells. Finally the dilution of complement required in 50 μl for a standard excess of 4 $CH_{50}$ units is calculated.

3.3.5 *Technical notes*

(i) Guinea pig complement can be purchased in preserved form (see Appendix for suppliers). It should be reconstituted and stored according to the supplier's recommendation. Normally it is used without further dilution.

(ii) Fresh guinea pig serum is harvested from clotted blood withdrawn by cardiac puncture (see Chapter 2, Section 5.2.3) under sterile conditions, ensuring minimum haemolysis. The serum is carefully separated from the clot as soon as possible and spun at 200 *g* to remove any red cells. The complement will remain active in serum stored in small aliquots at −70°C.

(iii) In the titration, the starting dilution of complement should be such as to provide in tube 2 a minimum of lysis. In practice a dilution of 1:30 (in HD) made from the 1:3 absorbed complement is near optimal but this should be adjusted according to trial result.

### 3.4 **Haemolytic antibody assay**

3.4.1 *Comparison of haemolytic and haemagglutination titres of sera*

(i) Mark out a microtitration plate as in Section 2. Add 50 μl of HD to each well.

(ii) With a diluting tulip (50 μl) dilute normal rabbit serum in row A and test sera in duplicate rows beneath.

(iii) Add to one row of each test serum a further 50 μl of HD (replaces complement) and to all other rows 50 μl of 4 $CH_{50}$ units of complement as determined in Section 3.3.

(iv) Add to all wells 50 μl of 1% SRBC suspension.

(v) Mix adequately by tapping the plate, cover and incubate at 37°C for 1−2 h. To examine lytic reactions view the plate over a white background. The haemolytic titre is defined as the last dilution giving total lysis (i.e. no evidence of intact red cells settling in the well). V-bottomed wells give more accurate results. Row H (nil serum control) should contain no lysis or agglutination. Row A (normal serum control) may show agglutination and lysis (partial) in the first well(s) due to heterophile antibodies.

#### 3.4.2 *Comparison of IgM and IgG haemolytic titres using 2-ME*

Follow procedures (i) and (ii) from Section 3.4.1.

(iii) Add to one row of each test serum a further 50 μl of HD (replaces 2-ME) and to all other rows (in fume cupboard) 50 μl of 2-ME. Cover the plate and stand it on the bench for 30 min.

(iv) To all wells add 50 μl of complement (4 $CH_{50}$) followed by 50 μl of 1% SRBC suspension. Mix, cover and incubate as before.

(v) Interpretation: *Figure 2*, rows D and E, shows a haemolytic assay, with (E) and without (D) 2-ME pre-treatment of the serum (in this case using 2% SRBC). The effect of the 2-ME is to reduce the haemolytic titre by three serial dilutions (i.e. 8-fold).

## 4. ANTIGLOBULIN (COOMBS) ASSAYS

### 4.1 **Assays for non-agglutinating antibodies**

#### 4.1.1 *Principle*

IgG antibodies to cell surface antigens which are exposed at low density, or are not well displayed, sometimes behave like monovalent antibodies; presumably because they are unable to form stable crosslinkages. Rhesus antibodies of IgG class are of this type. The binding of antibody may be readily revealed, however, by applying a second stage antiglobulin reagent. It is, of course, necessary to wash away unbound immunoglobulins present in the first stage antiserum which would otherwise compete for the antiglobulin.

#### 4.1.2 *Special equipment and reagents*

Low speed centrifuge with microtitration plate holders.
Pasteur pipettes with drawn out fine tips.
Antiglobulin reagents: anti-IgM, anti-IgG, or mixed specificity, absorbed versus SRBC [see Section 3.3.3(i)].

#### 4.1.3 *Method*

The titration may be performed in plates or tubes. The assay as applied to the detection of rhesus antibody by the tube method is described in Volume II. If plates are used, a special carrier is required for spinning (only usable on some centrifuges). Low speed centrifugation (just fast enough to pellet the red cells), of short duration is all that is necessary. The supernatant is then sucked off with a drawn out pipette. The cells are washed once with PBS, again centrifuged and finally resuspended in 100–200 μl of dilute (~1:100) antiglobulin by gently squirting back and forth along the rows starting at the weak end. After the cells have settled the end point is read.

### 4.2 **Antiglobulin titration using pre-sensitized cells**

#### 4.2.1 *Principle*

A simple and useful way of titrating antiglobulin reagents (anti-IgM or anti-IgG, or mixed specificity), as required for instance to standardize the Coombs test or to compare antiglobulin serum titres in quality control, is to use red cells pre-coated with a sub-agglutinating dose of anti-erythrocyte antibodies. IgM and IgG antibody components can be separated by chromatography as described in Chapter 2, Section 3.2.3 and used independently to sensitize the cells.

4.2.2 *Method*

The cells can be sensitized with antibody as described in Section 3.3.2. Using three concentrations of antibody to coat the cells, a dose can usually be found at the first attempt that gives a very sensitive antiglobulin test without any tendency for spontaneous agglutination. The assays should, however, be performed in buffer containing 2% heat-inactivated fetal calf serum (FCS) as red cells coated at the optimal level for antiglobulin assays of this type tend to clump in protein-free buffer.

The test is performed in the same way as direct haemagglutinations but it may be necessary to titrate antiglobulins over a 24-well series to reach an end point, or to start the titrations at 1:1000 in view of the sensitivity of the system. Sensitized cells (0.5%) can be used for maximal sensitivity.

The following controls are necessary:

(i) Antibody-coated cells in buffer.
(ii) Normal serum of antiglobulin species with coated cells.
(iii) Antiglobulin with normal cells.

4.2.3 *Other applications*

The assay can be applied to testing the purity of IgM and IgG red cell-sensitizing antibodies for use in raising antiglobulin reagents (see Chapter 2, Section 3.2.3). In passive tests with antigen-coupled red cells, isotype-specific antiglobulins can be used to examine the isotype profile of a first stage non-agglutinating antibody, or other antibodies diluted to a sub-agglutinating level.

## 5. PASSIVE HAEMAGGLUTINATION (PHA) USING ANTIGEN-COATED RED CELLS

### 5.1 **Comparison of antigen-coupling methods**

One of the earliest agents used to couple antigens to red cells was *bis*-diazotized benzidine (BDB). The use of this and other early reagents is reviewed in ref. 6. BDB is a molecule with two reactive diazonium groups which, ideally, would couple to a cell surface tyrosyl or imidazyl group and the other to a similar group on the soluble antigen. A 'one-stage' coupling agent, such as BDB, is added directly to a mixture of red cells and antigen and there is no way of restricting the formation of undesirable antigen–antigen or erythrocyte–erythrocyte bonds or of preventing modification of the antigen by the coupling agent.

The first satisfactory two-stage reagent, tannic acid, was developed by Boyden (7). When red cells are exposed to vegetable tannins (polyphenyl glycosides), under protein-free conditions, part of the molecule attaches to red cell surface protein but other protein-binding groups remain available for attachment, after a light wash, to a soluble protein antigen. Since all the available coupling agent is attached to red cells by the time the protein antigen is added, coupling is more efficient and little or no modification of the antigen occurs. This method is, in principle, therefore, more satisfactory than a 'one-stage' technique. It was further refined by Stavitsky (8).

Many other crosslinking reagents, such as difluorodinitrobenzene (6) and diisocyanate (9) have been used and one that has gained considerable popularity is glutaric dialdehyde or glutaraldehyde as it has come to be known. Just as in the BDB technique, a standard amount of coupling agent is added to a mixture of washed red cells and protein antigen (10,11). Coupling is rapid and is complete after a few minutes; moreover glutaraldehyde and the coated red cells are more stable than native red cells (in contrast to the BDB technique) and will keep for some weeks at 4°C. However, the protein antigen may be seriously modified by the tyrosyl, imidazyl, sulphydryl and amino groups having reacted with the aldehyde. It is not possible to use glutaraldehyde as a two-stage reagent because the two active groups readily combine with groups on a single red cell and the reagent, being lipophilic, readily penetrates the red cell (unlike formaldehyde) and fixes protein inside the cell (12). Proteins may, however, be directly attached to red cells which have been previously fixed with formaldehyde, glutaraldehyde or other aldehydes and washed (12,13). This technique requires nothing more than incubation of the fixed red cells with small amounts of antigen. Once coated the red cells retain the antigen indefinitely, even after repeated washing. The coated cells may be stored in azide and used over long periods. Another method which depends upon aldehyde groups for coupling is based on mild periodate oxidation of carbohydrate groups on the surface of red cells, followed by adding the protein antigen to the red cells bearing the newly generated reactive groups (14).

A simple direct 'one-stage' coupling agent, chromic chloride, first introduced by Jandl and Simmons (15) has become very widely used since its re-introduction by Gold and Fudenberg (16). Apart from its simplicity, its advantages are that virtually no modification of coupled protein occurs and a higher level of coating is possible (controlled by the amount of chromic chloride used) than with direct attachment to preserved cells (17). The method is so mild that even coupled antibodies retain activity whereas they are destroyed by glutaraldehyde coupling. Staphylococcal protein A and lectins (provided they do not react with sugar residues found on red cells) also couple satisfactorily and retain activity. Provided the right animal is chosen as a source of red cells and sterile precautions are observed throughout, the coated cells may be kept free of lysis when stored at 4°C for several weeks and remain active in agglutination tests.

## 5.2 Coupling protein antigens to red cells using chromic chloride

### 5.2.1 *Special materials and reagents*

Sterile capped 75 × 13 mm glass tubes.
Vortex mixer.
Dialysis equipment.
Small sterile capped tubes.
Hepes-RPMI medium (H-RPMI) with 2% heat-inactivated FCS for some uses (H-RPMI−2%FCS).
Chromic chloride: 1 mg/ml stock solution in saline, pH 5, matured for 1 month or more.
Sterile saline.
Protein antigens, at least 3 mg/ml.

5.2.2 *Method*

A recommended procedure (based on ref. 18) is as follows.

(i) Collect blood aseptically in an equal volume of sterile Alsever's solution. Observe sterile procedures throughout.

(ii) Mix the sheep blood to suspend the cells and transfer a sample into a sterile capped tube. Add sterile saline, replace the cap, mix and spin at 700 *g* on a bench centrifuge for 10 min. Suck off the supernatant and buffy coat and discard. Wash a further five times in sterile saline (700 *g*, 5 min), removing as much supernatant and buffy coat as possible each time. No trace of lysis should be present. Finally resuspend the cells in saline again and spin for 10 min at 1100 *g* to pack the cells.

(iii) Remove the supernatant completely, flick the tube to disperse and mix the cells. Suck up a fixed volume in a graduated pipette (2 ml or 5 ml) and transfer to a glass Universal containing 9 vol of sterile saline (10% suspension). Mix well and transfer 1 ml aliquots into labelled, sterile, capped 75 × 13 mm glass tubes.

(iv) When ready to proceed with the coatings, spin down the red cells and remove the supernatant. Add a measured volume of the solution of the protein to be coated to all the tubes over a trial range of antigen amount. The concentrations should be 1−3 mg/ml and the amounts added in the range 0.05−0.3 mg. 0.3 mg is an excess and usually gives results slightly superior to 0.1 mg; 0.05 mg is approximately the lower trial limit.

(v) Move the rack of tubes adjacent to a vortex mixer. Add the required amount of chromic chloride [0.4−1.2 ml—see Section 5.2.3(v)], using the mixer to ensure prompt mixing, tube by tube, as it is added. Wash down the walls of each tube by gently running in 1 ml of sterile saline.

(vi) Leave the capped tubes undisturbed overnight. Next day add 2 ml of H-RPMI medium to each tube and mix. Spin down the red cells, remove the supernatant and resuspend the cells in 4 ml of H-RPMI. Store at 4°C. Dilute 1:5 in H-RPMI−2% FCS before use in titrations.

5.2.3 *Technical notes*

(i) Proteins for coupling may be prepared by conventional salt fractionation and chromatographic methods but must be exhaustively dialysed against saline before use. The solution is then subjected to high speed centrifugation (11 000 *g*) or sterile filtration through a Millipore filter and the protein concentration calculated from an $A_{280}$ measurement. Small aliquots are measured out in small sterile tubes which are then stored in the cold (usually at −20°C). Since all proteins are coupled indiscriminately it is obviously desirable to use a purified protein for coupling. Absolute purity is, however, usually unnecessary.

(ii) Chromic chloride. A stock solution is prepared by dissolving 500 mg of $CrCl_3$. $6H_2O$ in 500 ml of saline. The $CrCl_3$ gradually hydrolyses with the formation of complex chromium ions and the pH falls. Every week for several weeks the pH is adjusted to approximately 5.0 (external indicator) by addition of drops of 1 M NaOH. Neutralization is accompanied by a colour change from a very pale blue to light green. After about a month of storage and neutralization, the pH remains relatively constant, requiring only the occasional drop of 1 M NaOH to neutralize. It is then considered 'mature' and ready for use. For use in the coating procedure the chromic chloride is diluted 1:10 in saline to 0.1 mg/ml.

(iii) Saline. It is imperative that phosphates, acetate, proteins and other extraneous materials should be eliminated from the solutions as they inhibit coupling. The red cells, the chromic chloride and the protein antigens must all be made up in saline. Sterile saline may be prepared by autoclaving in a hard glass bottle (soft glass releases alkali) or by membrane filtration.

(iv) Most proteins for coupling are best aliquoted and stored at −20 or −70°C. However some proteins (e.g. IgM) are denatured by freezing and are best stored at 4°C.

(v) The amount of chromic chloride needed for coupling is usually characteristic for a given protein preparation. On the first occasion a range of chromic chloride volumes (0.4−1.2 ml) should be tried and the highest dose giving a negative 'settled pattern' selected. Similar volumes should be required on future occasions.

## 5.3 Passive haemagglutination test with protein antigen-coupled red cells

### 5.3.1 *Principle*

The passive test is performed like a direct haemagglutination assay with serial dilution of antibody in rows of microtitration plates, addition of coated cells and the 'settled pattern' end point read after several hours. A second antiglobulin stage may be included following the protocol in Section 4.1.

### 5.3.2 *Example*

(i) Serially dilute a mouse monoclonal antibody (Mab) to human immunoglobulin $\kappa$ chains in 50 $\mu$l volumes of H-RPMI−2% FCS.

(ii) Add 50 $\mu$l of 0.5% (v/v) antigen-coupled SRBC (IgG $\kappa$ or free $\kappa$ light chains as antigen) to each antibody well and to control wells containing 50 $\mu$l of diluent alone.

(iii) After 2 h read the direct end point titre (first stage titre) for agglutinating antibodies.

(iv) Centrifuge the plate lightly (Section 4.1), remove the supernatant and wash the cells with 100 $\mu$l of diluent and centrifuge again. After removal of the wash supernatants, add 100 $\mu$l of sheep anti-mouse immunoglobulin reagent, at 1:100 dilution in diluent, to each well and resuspend the cells by pipette, starting at the weak end of each row. Read the end point of the indirect (second stage) titration 2 h later.

### 5.3.3 *Interpretation*

A positive result in a direct titration demands of the antibody the capacity to crosslink as well as to bind, whereas the second stage picks up antibodies capable of binding firmly to antigen but not effectively crosslinking. The two-stage titration is thus approximately equivalent to other 'binding' assays such as the enzyme-linked immunosorbent assay (ELISA) or the radiometric assay except that the antigen used remains in its native state (i.e. it is not denatured by attachment to a plastic surface).

5.3.4 *Technical notes*

(i) Because of the high sensitivity of the assay two rows may be needed to titrate out an antiserum, or an initial dilution of the antiserum may be made.

(ii) Antigen-coupled red cells tend to agglutinate spontaneously and require a protein-rich medium to limit this tendency. FCS is ideal as it contains no immunoglobulin. The source of the FCS should be checked, however, to ensure its fetal origin and it should be heat-inactivated to destroy complement.

(iii) When using an antiglobulin second stage titration it is essential to ensure that this contains no anti-red cell antibodies and, in the case of coupled immunoglobulin antigens, that there is no reactivity with the coupled species molecules.

(iv) Control wells should include:
- (1) diluent with antigen-coupled cells;
- (2) positive sera with uncoated cells;
- (3) normal sera with coated cells;
- (4) antiglobulin reagent with coated cells (where applicable).

### 5.4 **Passive haemagglutination with red cells coated with polysaccharides and haptens**

Polysaccharides in general may not be coupled to red cells by agents like chromic chloride or tannic acid and most do not attach spontaneously. There are two ways in which they may be coupled (see also Chapter 2, Section 3.1.2).

(i) Periodate oxidation of some of the sugar groups with the formation of aldehydes which link covalently to amino groups on the surface of the red cells (19,20).

(ii) Attachment of a long chain fatty acid, such as stearic acid, to the polysaccharide. The lipophilic part of the molecule becomes firmly embedded in the red cell membrane when derivatives of this kind are incubated at 37°C with red cells and the attached polysaccharide is displayed on the surface (21).

There are also, in fact, many natural lipopolysaccharide (LPS) substances of this kind which are produced by bacteria (e.g. LPS of Gram-negative organisms, teichoic acid derivatives shed by streptococci and staphylococci), and coating of red cells with these antigens requires nothing more than incubation of red cells with culture filtrate or a bacterial extract. Many antibiotics (e.g. penicillin) will also attach spontaneously. The uptake is an active process and requires an incubation period of 30–60 min at 37°C.

Haptenic groups may be introduced onto a red cell surface by simply incubating with a reactive derivative (e.g. fluorodinitrobenzene for the dinitrophenyl–hapten). An alternative method is the attachment of haptens via an LPS unit (22,23).

## 6. REVERSE PASSIVE HAEMAGGLUTINATION ASSAYS WITH ANTIBODY-COATED RED CELLS (RPHA)

### 6.1 **Principle**

Antigens can be detected and quantitated with great sensitivity by using antibody-coupled red cells with free combining sites which are agglutinated by antigen in solution. For agglutination to occur directly on antigen exposure to the red cells, the antigen must be sufficiently complex to present epitopes to antibodies on different cells and in the case of a single coupled Mab the antigen must present two identical epitopes spatially separated.

### 6.2 **Antibody coupling**

Immunoglobulin for coupling must be relatively pure and should ideally have the majority of molecules as specific antibody. This is no problem using Mabs purified from ascitic fluid, but polyclonal reagents must be either affinity-purified from an antiserum or, at the least, substantially purified as an immunoglobulin fraction of a potent reagent by ion-exchange chromatography (see Chapter 2, Section 10). Coupling to the red cells must be by a non-destructive method since the specific antigen-binding capacity of antibody is very labile. Two simple techniques in common use are described below.

#### 6.2.1 *Direct antibody attachment*

(i) Wash 0.1 ml aliquots of glutaraldehyde- or pyruvic aldehyde-fixed SRBC and resuspend in 2 ml of 0.1 M acetate buffer, pH 5.0 containing 10–100 μg of antibody.

(ii) After incubation at 37°C for 2 h with occasional mixing, wash the cells twice and resuspend in 2 ml of PBS containing 0.1% azide and 0.1% bovine serum albumin (BSA).

#### 6.2.2 *Chromic chloride*

The technique is exactly as described for protein antigens and it is the method of choice when maximal sensitivity is required. Red cells fixed with dimethyl suberimidate may also be coated by the chromic chloride technique (17) but this modification offers few advantages. It may be used to measure antigens in non-ionic detergent extracts of cells, but provided excess detergent is removed by resin treatment, conventional unfixed antibody-coated cells may be used (25).

### 6.3 **Interpretation**

In titrations of antigens by reverse PHA it is common to find a strong prozone at high concentrations of antigen. This may sometimes be so pronounced that only a few wells in the mid-titration zone are positive and this is especially likely to occur if the coated antibody is of low potency or has not been adequately purified.

### 6.4 **Technical notes**

(i) In the single stage test, antigen is diluted out in wells of H-RPMI–2% FCS and then antibody-coated red cells added and the plate incubated. The end point titre of antigen is the final dilution giving positive agglutination of the cells. The test is useful both for detection of antigen and measurement of test antigen against a standard dilution reaction.

(ii) It is possible to add a second stage to the titration just as in the case of conventional PHA—by simply washing the cells and adding a second stage antibody into the suspension. This second stage antibody must obviously be directed against determinants on the antigen which are still available after it has been 'captured' by the red cell-bound antibody. If the first and second stage reagents are both Mabs they must be directed against spatially distant epitopes.

(iii) Another useful device, especially when dealing with monovalent antigens, is to use a mixture of antibody-coated cells for the titration, each with specificity for non-proximal determinants (24). Sometimes this is best performed in two stages, for example; red cell-bound anti-free λ chain reagent may be used at stage one to pick up free λ chains from a mixture of free and heavy chain-bound λ chains. After washing, the cells are resuspended and mixed with red cells coated with an anti-general λ antibody.

## 7. HAEMAGGLUTINATION INHIBITION (HAI) ASSAYS

### 7.1 **Principle**

Agglutination of antigen-coated red cells by antibody at a concentration marginally above its minimum agglutinating dose (MAD) is easily inhibited by very small quantities of the same antigen in solution. Setting up the test system involves a first titration of the antibody by serial dilution to determine the approximate MAD, and then a second titration starting at an estimated 4 MAD to find the accurate end point. A trial inhibition with standard antigen is then performed using 2 MAD units of antibody, the aim being to achieve a sensitive inhibition, expressed as no agglutination with antigen diluted to trace amounts, followed by agglutination wells. The transition represents the antigen end point. Samples can then be tested for presence of antigen in solution and the antigen quantitated by dilution, comparing the end point with that of the standard.

### 7.2 **Method**

(i) Prepare antigen-coated red cells as for PHA tests (Section 5.2).

(ii) Perform the two preliminary antibody agglutination assays (approximate and accurate end points) using 0.5% antigen-coated red cells and H-RPMI−2% FCS. From the second assay determine the required antibody dilution for 2 MAD with 0.5% cells and prepare a volume of antibody at this dilution.

(iii) Standard antigen inhibition. Add 50 μl of diluent to wells of three rows of a plate. With a tulip add 50 μl of antigen to the first well of row A and serially dilute over two rows. If the starting antigen concentration is 1 mg/ml the dilution will extend to less than 100 pg/ml. Add 50 μl of antigen-coated cells to the wells, mix, cover and read results from the settled pattern after 1−2 h. The third row of wells is used for controls. These should include coated cells with antigen (no antibody) and coated cells with antibody alone.

(iv) Test inhibitions. Dilute out standard and test antigens in parallel on the plate. The range of standard antigen can be restricted to give an end point within a single row. Test antigens may still require two rows to find the end point. Controls should include test antigens with coated and normal cells and coated cells with antibody alone.

### 7.3 Technical notes

(i) HAI is often preferred to reverse PHA as a method for measurement of antigen. It has some advantages, for example antibody does not have to be purified from the antiserum or ascitic fluid; the antigen may be monomeric (i.e. without duplicated epitopes) and is tested in its native state.

(ii) Excellent results can sometimes be obtained by this method. When it fails it is usually because the antibody is of low affinity and will then combine preferentially with the repeating display of antigenic determinants on the red cell surface and not with free antigen.

(iii) A useful modification of HAI for the study of allotypic markers of animal immunoglobulins combines the principles of an antiglobulin test and inhibition. Antibodies to SRBC can be used to sensitize the cells at a sub-agglutinating dose so that anti-allotype sera achieve the agglutination. This can be inhibited by free allotype. Animals homozygous for known reference allotypes can be readily immunized with red cells to prepare the sensitizing allotypic antibody. This strategy has been applied successfully to the typing of rabbit allotypic loci in non-precipitating systems. In one method non-agglutinating isoantibodies to rabbit red cell allelic antigens have been used for rabbit red cell sensitization (26).

## 8. THE COMPLEMENT FIXATION TEST (CFT)

### 8.1 Principle

The formation of antigen–antibody complexes *in vitro* can be determined by measurement of the fixation (consumption) of complement added in standard, limiting amount from an extrinsic source. Intrinsic complement from test sera is inactivated in a preliminary heating step. The test critically depends upon the presence of complement-fixing antibody isotypes but can be used to detect both antibody and antigen.

### 8.2 General method

Standard antigen and complement-fixing reference antibody (inactivated) are diluted out as mixtures in a chequerboard titration of antigen versus antibody. A limiting amount of active complement is added to the wells; where complement is fixed it is no longer available to mediate complement-dependent lysis of antibody-coated cells when these are added to the wells as an indicator system of complement availability. That part of the chequerboard that shows a sensitive transition from no lysis to complete lysis reveals the conditions of antigen and antibody concentration that can be recreated in test conditions for determining antigen or antibody.

The steps in setting up the assay are as follows.

(i) Preparing a haemolytic indicator system of antibody-sensitized red cells and a measured complement dose. For antibody, the use of 6 MAHD units is recommended and for complement 4 $CH_{50}$ units. The titrations required to determine these antibody and complement levels, and the preparation of sensitized red cells and absorbed complement, are described in Sections 3.3.1–3.3.4.

(ii) Performing the antigen–antibody chequerboard dilution assay, with added complement.

(iii) Adding the haemolytic indicator and determining the sensitive end point zone for use in the test system.

## 8.3 Detailed methods

### 8.3.1 *Chequerboard titrations*

(i) Heat the antiserum at 56°C for 30 min to inactivate any intrinsic complement. To all wells of a microtitration plate add 50 μl of HD (see Section 3.2). Then double dilute the antibody in parallel across all eight rows (A–H) to column 11 (1:2048). Column 12 is left as antibody-free (antigen-alone) control.

(ii) In seven 2 ml glass tubes, labelled B–H, prepare in each a 1 ml dilution of antigen in HD, B–H being a 2-fold dilution series. Place 50 μl of dilution B (highest concentration) in each well of row B; repeat for C–H. Row A is left as an antigen-free (antibody alone) control.

(iii) Add 50 μl of absorbed complement, containing 4 $CH_{50}$ units, to each well of the plate, mix, cover and stand at 4°C overnight (16 h).

### 8.3.2 *Lytic indicator step*

The next morning incubate the plate at 37°C for 30 min and then add 50 μl of 0.5% antibody-sensitized (6 MAHD) red cells, mix, cover and re-incubate at 37°C for 60 min or until haemolysis develops.

### 8.3.3 *Interpretation*

With suitable dilutions of antigen and antibody in the chequerboard, the zone at the top left of the plate with high reagent concentrations will have fixed most complement, leading to zero or minimal lysis. Reagent conditions may need to be adjusted to achieve this result.

A high level of haemolysis should be observed in the antigen alone and antibody alone control wells (column 12 and row A) and in mixed wells where antigen and antibody become critically too dilute to allow much formation of complement-fixed complexes (downwards and to the right respectively). The centre region of the plate should sustain complement fixation with a zone of wells showing minimum lysis. This region offers suitable proportions of antigen and antibody for subsequent testing of unknowns under the most sensitive conditions.

### 8.3.4 *Testing sera or antigens*

This can now be done in single rows using either standard antigen concentration or standard antibody dilution and varying the other reagent, the conditions being selected from the chequerboard result.

## 8.4 Technical notes

(i) The final step of obtaining a settled pattern of cells, to determine haemolysis, may be shortened by gentle centrifugation of the plates.

(ii) Control wells of antigen alone and antibody alone should both provide total haemolysis as the complement is unused. Failure of both to provide this indicates inadequate antibody-coating of the indicator cells. Failure of one control indicates anti-complementary activity in the added reagent, which should be replaced.

(iii) Because IgM has a higher haemolytic efficiency, this antibody class is preferentially detected. For this reason the complement fixation test remains popular as a clinical test of recently acquired infections with both bacteria and viruses.

(iv) The most accurate method for measuring the test is to use $^{51}$Cr-labelled, antibody-coated cells. A 50% suspension of cells is labelled for 30 min at 22°C, washed three times in HD and diluted to 0.5%. The test is performed as in Sections 8.3.1 and 8.3.2 but after incubation with the indicator cells the plate is cooled to 4°C and centrifuged at 2000 *g* for 5 min. Supernatant (75 μl) of each well is removed for counting in a gamma counter. An alternative, after spinning, is to measure the absorption of haemoglobin of the supernatant at 413 nm by spectrophotometry. In both methods complement fixation is compared with controls of 100% lysis and results expressed as percentage complement fixed. The assay end point by these methods can be determined as the highest titre of antibody, or antigen dilution, that gives greater than 50% complement fixation (27).

## 9. REFERENCES

1. Amman,A.J., Borg,D., Kondo,O.L. and Wara,D.W. (1977) *J. Immunol. Methods,* **17**, 365.
2. Chao,W. and Yokoyam,M.T. (1977) *Clin. Chim. Acta,* **78**, 79.
3. Singer,J.M. and Plotz,C.M. (1956) *Am. J. Med.,* **21**, 888.
4. Winblad,S. (1960) *Acta Pathol. Microbiol. Scand.,* **49**, 500.
5. Gleeson-White,M.H., Heard,D.H., Mynors,L.S. and Coombs,R.R.A. (1950) *Br. J. Exp. Pathol.,* **31**, 321.
6. Ling,N.R. (1961) *Immunology,* **4**, 49.
7. Stephen,V. and Boyden,S.V. (1951) *J. Exp. Med.,* **93**, 107.
8. Stavitsky,A.B. (1954) *J. Immunol.,* **72**, 368.
9. Gyenes,L. and Sehon,A.H. (1964) *Immunochemistry,* **1**, 43.
10. Onkelinx,E., Mendelmans,W., Jonian,M. and Loutie,R. (1969) *Immunology,* **16**, 35.
11. Avramaeus,S., Tan Don,B. and Chinlon,S. (1969) *Immunochemistry,* **6**, 67.
12. Ling,N.R. (1961) *Br. J. Haematol.,* **7**, 299.
13. Hirata,A.A., Emirick,A.J. and Boley,W.F. (1973) *Proc. Soc. Exp. Biol. Med.,* **143**, 761.
14. Sanderson,C.J. and Wilson,D.V. (1971) *Immunochemistry,* **8**, 163.
15. Jandl,J.H. and Simmons,R.L. (1957) *Br. J. Haematol.,* **3**, 19.
16. Gold,E.R. and Fudenberg,H.H. (1967) *J. Immunol.,* **99**, 859.
17. Ling,N.R., Stephens,G., Bratt,P., and Dhaliwal,H.S. (1979) *Mol. Immunol.,* **16**, 637.
18. Ling,N.R., Bishop,S. and Jefferis,R. (1977) *J. Immunol. Methods,* **15**, 279.
19. Bankert,R.B., Mayers,G.L. and Pressman,D. (1977) *J. Immunol.,* **118**, 1265.
20. Ghanta,V.K., Hamlin,N.M., Pretlow,T.G. and Hiramoto,R.N. (1972) *J. Immunol.,* **109**, 810.
21. Hammerling,U. and Westphal,O. (1967) *Eur. J. Biochem.,* **1**, 1.
22. Bankert,R.B., Mayers,G.L. and Pressman,D. (1979) *J. Immunol.,* **123**, 2466.
23. Bankert,R.B. and Wolf,B. (1974) *J. Immunol.,* **112**, 1782.
24. Ling,N.R., Lowe,J., Hardie,D., Evans,S. and Jefferis,R. (1983) *Clin. Exp. Immunol.,* **52**, 234.
25. Parish,C.R., O'Neill,N. and McKenzie,I.F.C. (1980) *J. Immunol. Methods,* **39**, 223.
26. Mandy,W.J. and Todd,C.W. (1969) *Immunochemistry,* **6**, 811.
27. Brown,D. and Hobart,M.J. (1981) In *Techniques in Clinical Immunology,* Thompson,R.A. (ed.), Blackwell Scientific Publications, Oxford, 2nd edition, Chapter 4, p. 80.

APPENDIX

## Suppliers of Specialist Items

EQUIPMENT AND DISPOSABLES

Bench centrifuge (with plate carriers): C.411 model with carriers 11.17.41.68—*S.A.Jouan, 130 Western Road, Tring, Herts HP23 4BU, UK*

$CO_2$ incubator: *Flow Laboratories, Woodcock Hill, Harefield Road, Rickmansworth, Herts WD3 1PQ, UK*

Cork borers: *Gallenkamp Co. Ltd, Technico House, Christopher Street, London EC2P 2ER, UK*

Dialysis tubing: *Medicell International Ltd, 239 Liverpool Road, London N1, UK*

Diluting tulips: *Flow Laboratories, Woodcock Hill, Harefield Road, Rickmansworth, Herts WD3 1PQ, UK*

Electrophoresis box with large buffer reservoirs: *Chemical Laboratory (Chemlab) Instruments Ltd, 129 Upminster Road, Hornchurch, Essex RM11 3XJ, UK*

Electrophoresis units: *Paines and Byrne Ltd, Pabyrn Laboratories, 177/179 Bilton Road, Perivale, Greenford, Middlesex UB6 7HG, UK*

Film developer and fixer: *Kodak D19, Eastman-Kodak Co., PO Box 10, Dallimore Road, Manchester M23 9NJ, UK*

Gelbond$^{TM}$ film: *Marine Colloids Division, FMC Corporation Bioproducts, Rockland, ME 04841, USA*

Gel plate storage boxes (10×10×1.5 cm): *Sterilin Ltd, Lampton House, Lampton Road, Hounslow, Middlesex TW3 4EE, UK*

Gel punches (and punch assemblies): *Shandon Southern Products Ltd, Chadwick Road, Astmoor, Runcorn, Cheshire WA7 1PR, UK*

Glass agar plates (8×8 cm): *Appleton Woods Ltd, 313 Heeley Road, Selly Oak, Birmingham B29 6EN, UK*

Hamilton syringes: *FSA Laboratory Supplies, Bishop Meadow Road, Loughborough, Leics LE11 0RG, UK*

Hard-glass bottles (Duran bottles; graduated, screwcap, autoclavable): *FSA Laboratory Supplies, Bishop Meadow Road, Loughborough, Leics LE11 0RG, UK*

Homogenizer for cells; motor-driven: Ultra Turrax Model T25 and Griffiths tubes. *BDH Apparatus Division, PO Box 8, Dagenham, Essex RM8 1RY, UK*

IEP gel cutters: *Shandon Southern Products Ltd, Chadwick Road, Astmoor, Runcorn, Cheshire WA7 1PR, UK*

Intensifier screens: DuPont Cronex 'Lightning Plus' (with Agfa Curix RP-2 films): *DuPont Instruments, Pecks Lane, Newtown, CT 06470, USA*

Inverted microscope; Olympus CK with x10 objective and binocular wide field x15 eyepieces: *Olympus Optical Co., 43-2 Hatagaya 2 Chome, Shibuya-Ku, Tokyo, Japan*

Laminar flow tissue culture hood: *Flow Laboratories, Woodcock Hill, Harefield Road, Rickmansworth, Herts WD3 1PQ, UK*

Liquid nitrogen containers: *Cryoservices, Blackpole Trading Estate, Blackpole Road, Worcester WR3 8SG, UK*

Liquid nitrogen polypropylene storage vials: *Flow Laboratories, Woodcock Hill, Harefield Road, Rickmansworth, Herts WD3 1PQ, UK*

Micropipettes (Finnpipettes): *Jencon Scientific Ltd, Cherry Court Way, Industrial Estate, Leighton Buzzard, Bedford LU7 8UA, UK*

(Gilson): *Anachem Ltd, Anachem House, 20 Charles Street, Luton, Beds LU2 0EB, UK*

(fixed and variable types): *Oxford Labs, Boehring Corp. (London), Boehring Mannheim House, Bell Lane, Lewes, East Sussex BN7 1LG, UK*

Microtitration plates (U- and V-bottom): *Flow Laboratories, Woodcock Hill, Harefield Road, Rickmansworth, Herts WD3 1PQ, UK*

Multichannel micropipettes (Titertek): *Flow Laboratories, Woodcock Hill, Harefield Road, Rickmansworth, Herts WD3 1PQ, UK and Dynatech Laboratories Ltd, Daux Road, Billingshurst, Sussex RH9 5J, UK*

Parafilm: *American Can Company, Greenwich, CT, USA*

Petri dishes; 5 cm sterile tissue culture grade: *Sterilin Ltd, Lampton House, Lampton Road, Hounslow, Middlesex TW3 4EE, UK*

P1-pump pipette aids: *FSA Laboratory Supplies, Bishop Meadow Road, Loughborough, Leics LE11 0RG, UK*

Polycarbonate centrifuge tubes: *Techmate Ltd, 10 Bridgeturn Avenue, Old Wolverton, Milton Keynes MK12 5QL, UK*

Power packs for electrophoresis (Vokam): *Shandon Southern Products Ltd, Chadwick Road, Astmoor, Runcorn, Cheshire WA7 1PR, UK*

Radial immunodiffusion plate reader: *Transidyne General Corp., Ann Arbor, MI, USA*

Radial immunodiffusion ring measuring scale: *Partigen-Behring Institute, Behring Diagnostics, Hoechst House, Salisbury Road, Hounslow, Middlesex TW4 6JH, UK*

Rotary cell mixer (Type BCM): *Voss of Maldon, Essex, UK*

SDS–PAGE apparatus: *Pharmacia/LKB Biotechnology, Pharmacia House, Midsummer Boulevard, Milton Keynes MK9 3HP, UK*

Minigel apparatus: *Biorad Laboratories Ltd, Caxton Way, Holywell Industrial Estate, Watford, Herts WD1 8RP, UK*

Shaker-incubator: *New Brunswick Scientific (UK) Ltd, 26–34 Emerald Street, London WC1N 3QA, UK*

Sterile capped glass tubes: *Flow Laboratories, Woodcock Hill, Harefield Road, Rickmansworth, Herts WD3 1PQ, UK*

Sterile filtration: *Sartorius Instruments Ltd, 18 Avenue Road, Belmont, Sutton, Surrey, UK*

Millipore filters (0.2 and 0.45 μm pore size): *Millipore (UK) Ltd, 11/15 Peterborough Road, Harrow, Middlesex HA1 2YH, UK*

Stopwatch: *FSA Laboratory Supplies, Bishop Meadow Road, Loughborough, Leics LE11 0RG, UK*

Teflon double-edged syringe adaptor for preparing water-in-oil emulsions: *Becton and Dickinson Ltd, Lab Products Division, Between Towns Road, Cowley, Oxford OX4 31Y, UK*

Test tubes, 30 ml Pyrex, round-bottomed: *FSA Laboratory Supplies, Bishop Meadow Road, Loughborough, Leics LE11 0RG, UK*

Tissue culture flat-bottomed flasks: *Sterilin Ltd, Lampton House, Lampton Road, Hounslow, Middlesex TW3 4EE, UK*

Tissue culture universal bottles: *Sterilin Ltd*

Tissue culture 24-well (2 ml) trays 'Linbro': *Flow Laboratories, Woodcock Hill, Harefield Road, Rickmansworth, Herts WD3 1PQ, UK*

Tissue culture 96-well plates 'Costar': *Northumbria Biologicals Ltd, South Nelson Industrial Estate, Cramlington, Northumberland NE23 9HL, UK*

Vortex mixer (whirlmixer): *FSA Laboratory Supplies, Bishop Meadow Road, Loughborough, Leics LE11 0RG, UK*

Warming plate; manufactured by Photax: *available at all major photographic suppliers*

Water bath (type JB1): *Grant Instruments Ltd, Barrington, Cambridge CB2 5Q1, UK*

Western blot (electroblot) and immunostaining apparatus: *Pharmacia/LKB Biotechnology, Pharmacia House, Midsummer Boulevard, Milton Keynes MK9 3HP, UK; Biorad Laboratories Ltd, Caxton Way, Holywell Industrial Estate, Watford, Herts WD1 8RP, UK*

Nitrocellulose membrane-Hybond C: *Amersham International plc, Lincoln Place, Green End, Aylesbury, Bucks HP29 2TP, UK*

X-Ray film and cassettes; Agfa Curix RP-1: *DuPont Cronex; DuPont Instruments, Peck's Lane, Newton, CT 06470, USA*

## MATERIALS, CHEMICALS AND REAGENTS

### Adjuvants

Arlacel A: *Sigma Chemical Co. Ltd, Fancy Road, Poole, Dorset BH17 7NG, UK*

Bayol F: *Esso Petroleum Co., Purfleet, Essex, UK*

*B.pertussis* vaccine: *Wellcome Diagnostics, Temple Hill, Dartford, Kent DA1 5AH, UK*

FCA and FIA: *Difco Laboratories Ltd, PO Box 14B, Central Avenue, East Molesey, Surrey KT8 0SE, UK*

FCA can be provided with *M.tuberculosis* (H37Ra) or with *M.butyricum*

Liposomes; Lecithin/Sphingomyelin standard solution: *Sigma Chemical Co. Ltd, Fancy Road, Poole, Dorset BH17 7NG, UK*

*M.bovis* BCG vaccine: *Glaxo Laboratories Ltd, Greenford, Essex, UK*

*M.tuberculosis*: *Ministry of Agriculture Veterinary Research Station, Weybridge, Surrey, UK and MRC Microbiological Supplies Division, Porton Down, Wiltshire, UK*

Muramyl dipeptide: *Sigma Chemical Co. Ltd, Fancy Road, Poole, Dorset BH17 7NG, UK*

**Antisera, monoclonal antibodies and antigen standards**
These suppliers include those offering HRP-antiglobulin conjugates for isotype screening.
Antiserum catalogue: *Linscott Directory of Immunological and Biological Reagents, PO Box 55, East Grinstead, Sussex RH19 3YL, UK*
Anti-SRBC (rabbit haemolytic and agglutinating reagents):
*Wellcome Diagnostics, Temple Hill, Dartford, Kent DA1 5AH, UK*
*BDS Biologicals Ltd, Vincent Drive, Edgbaston, Birmingham B15, UK*
*Behring Diagnostics, Hoechst UK Ltd, Hoechst House, Salisbury Road, Hounslow TW4 6JH, UK*
*Dako Ltd, 22 The Arcade, The Octagon, High Wycombe, Bucks HP11 2HT, UK*
*Miles Laboratories Ltd, PO Box 37, Stoke Court, Stoke Poges, Slough SL2 4LY, UK*
*Nordic Immunological Labs, PO Box 544, Maidenhead, Berks SL6 2PW, UK*
*Serotec Ltd, 22 Bankside Approach, Kidlington, Oxford OX5 1JE, UK*
*Oxoid Limited, Wade Road, Basingstoke, Hampshire RG24 0PN, UK*
WHO standard serum (for serum protein standards): *WHO International Reference Centre for Immunoglobulins, Institute of Biochemistry, University of Lausanne, Lausanne, Switzerland*

**Chromatography gels**
Cellulose carbonate: *Sigma Chemical Co. Ltd, Fancy Road, Poole, Dorset BH17 7NG, UK*
DE-52 (DEAE): *Whatman Biochemicals Ltd, Springfield Mill, Maidstone, Kent, UK*
Protein A–Sepharose-4B: *Pharmacia/LKB Biotechnology, Pharmacia House, Midsummer Boulevard, Milton Keynes MK9 3HP, UK*
Sephadex (G25, 200 etc.), Sephacryl (S200, 300): *Pharmacia/LKB Biotechnology, Pharmacia House, Midsummer Boulevard, Milton Keynes MK9 3HP, UK*
Sepharose 4B, 6B and CL-4B: *Pharmacia/LKB Biotechnology*

**Enzymes**
HRP, papain, pepsin, urease: *Sigma Chemical Co. Ltd, Fancy Road, Poole, Dorset BH17 7NG, UK*

**Hapten–protein conjugations**
(i) *Carriers*
Haemocyanin (Keyhole limpet [KLH] and Limulus): *Sigma Chemical Co. Ltd, Fancy Road, Poole, Dorset BH17 7NG, UK*
Species serum albumins and gamma globulins, porcine thyroglobulin: *Sigma Chemical Co. Ltd*
Ficoll: *Pharmacia/LKB Biotechnology, Pharmacia House, Midsummer Boulevard, Milton Keynes MK9 3HP, UK*
(ii) *Haptens*
DNP and TNBS: *Sigma Chemical Co. Ltd*
(iii) *Linking agents*
Glutaraldehyde; water-soluble carbodiimide; *N,N'-O*-phenylenedimaleimide; SPDP: *Sigma Chemical Co. Ltd*

Chromic chloride (chromium chloride, $CrCl_3$): *Sigma Chemical Co. Ltd*
CNBr: *Sigma Chemical Co. Ltd.*

**SDS–PAGE**

Gels, buffers, molecular weight markers, protein stains etc.: *Pharmacia/LKB Biotechnology, Pharmacia House, Midsummer Boulevard, Milton Keynes MK9 3HP, UK*
Silver staining kit: *Biorad Laboratories Ltd, Caxton Way, Holywell Industrial Estate, Watford, Herts WD1 8RP, UK*
Non-fat milk powder: St Ivel 5 pints (or similar brand): *Grocery Stores and Supermarkets*

**Tissue culture: cells, media and supplements**

B95-8 marmoset cell line: *from Dr J.Gordon, Department of Immunology, Medical School, University of Birmingham, Birmingham B15 2TJ, UK*
DMSO: *Sigma Chemical Co. Ltd, Fancy Road, Poole, Dorset BH17 7NG, UK*
Dulbecco's medium: *Oxoid Ltd, Wade Road, Basingstoke, Hampshire RG24 0PN, UK*
Dulbecco's A+B buffer: *Oxoid Ltd*
Fetal calf serum (FCS): *Gibco Europe Ltd, Unit 4, Cowley Mill Trading Estate, Longbridge Way, Uxbridge UB8 27G, UK*
Hepes: *BDH Ltd, Four Ways, Atherstone, Warwick CV9 1JQ, UK*
Hepes–RPMI medium (H–RPMI): *Gibco Europe Ltd, UK*
Horse serum: *Gibco Europe Ltd, UK*
Hypoxanthine: *BDH Ltd, UK*
2-Mercaptoethanol (2-ME): *Sigma Chemical Co. Ltd, Fancy Road, Poole, Dorset BH17 7NG, UK*
Methotrexate (aminopterin): *Lederle Laboratories Division, American Cyanamid Company, Pearls River, New York, USA*
Penicillin: *Gibco Europe Ltd, UK*
Plasmacytoma cell lines with HGPRT deficiency: P3-NS1/Ag4-1 and X63/Ag8.653 *Gift of Professor C.Milstein, MRC Laboratory of Molecular Biology, Hills Road, Cambridge CB2 2QH, UK*
(now carried by many laboratories—details from N.R.Ling)
Polyethylene glycol (PEG) 1500: *BDH Ltd, Four Ways, Atherstone, Warwick CV9 1JQ, UK*
RPMI-1640 tissue culture medium with L-glutamine: *Gibco Europe Ltd, UK*
Streptomycin: *Gibco Europe Ltd*
Thioguanine (anhydrous): *BDH Ltd, Four Ways, Atherstone, Warwick CV9 1JQ, UK*

**Other reagents and chemicals**

Agar (Difco Bacto Agar, Difco Special Agar Noble, Difco Oxoid Ion Agar): *Difco Laboratories Ltd, PO Box 14B, Central Avenue, East Molesey, Surrey KT8 0SE, UK*
Agarose: *Pharmacia/LKB Biotechnology, Pharmacia House, Midsummer Boulevard, Milton Keynes MK9 3HP, UK*
*BDH Chemicals Ltd, Poole, Dorset BH12 4NN, UK*
Bacto-agar: *Difco Laboratories Ltd*
Bromophenol Blue stain: *Sigma Chemical Co. Ltd, Fancy Road, Poole, Dorset BH17 7NG, UK*

Complement (preserved): *Wellcome Diagnostics, Temple Hill, Dartford DA1 5AH, UK*
Coomassie Brilliant Blue R stain: *Sigma Chemical Co. Ltd*
Cyanoacrylate tissue adhesive: (isobutyl 2-cyanoacrylate monomer-IBC-2):
Supplier: *Archibald Young & Son Ltd, 37 Constitution Street, Edinburgh EH6 7BG*
Manufacturer: *Ethicon Ltd, Edinburgh, UK*
Enzyme substrates—AEC and DAB: *Sigma Chemical Co. Ltd*
$H_2O_2$: *British Drug Houses (BDH Ltd)*
Ficoll-paque: *Pharmacia/LKB Biotechnology*
Iodoacetamide: *Sigma Chemical Co. Ltd*
KCNO (for carbamylation): *BDH Ltd*
Nonidet P-40 (NP-40): *Sigma Chemical Co. Ltd*
Phytohaemagglutinin: *Gibco Europe Ltd, Unit 4, Cowley Mill Trading Estate, Longbridge Way, Uxbridge UB8 27G,UK*
Phenylmethylsulphonyl fluoride (PMSF): *Sigma Chemical Co. Ltd*
Pristane: *Koch-Light Labs Ltd, Rockwood Way, Haverhill, Suffolk CB9 8PB, UK*
Sarcosyl: *Sigma Chemical Co. Ltd*
Silicone Repelcote: *BDH Ltd*
Sodium azide: *Sigma Chemical Co. Ltd*
Sodium deoxycholate: *Sigma Chemical Co. Ltd*
Trypan Blue: *Sigma Chemical Co. Ltd*
Tween 20 and Tween 40 (polyoxyethylene sorbitan monolaurate): *Sigma Chemical Co. Ltd*

## INDEX